Practical
MythTV

Building a PVR and Media Center PC

Stewart Smith
and Michael Still

Apress®

Practical MythTV: Building a PVR and Media Center PC

Copyright © 2007 by Stewart Smith and Michael Still

ISBN-13 (pbk): 978-1-59059-779-8

ISBN-10 (pbk): 1-59059-779-6

9 8 7 6 5 4 3 2 1

Lead Editor: Matt Wade
Technical Reviewer: Jay R. Ashworth
Editorial Board: Steve Anglin, Ewan Buckingham, Gary Cornell, Jason Gilmore, Jonathan Gennick, Jonathan Hassell, James Huddleston, Chris Mills, Matthew Moodie, Jeff Pepper, Dominic Shakeshaft, Jim Sumser, Matt Wade
Project Manager: Richard Dal Porto
Copy Edit Manager: Nicole Flores
Copy Editor: Kim Wimpsett
Assistant Production Director: Kari Brooks-Copony
Production Editor: Katie Stence
Compositor and Artist: Kinetic Publishing Services, LLC
Proofreader: Linda Seifert
Indexer: Ann Rogers
Cover Designer: Kurt Krames
Manufacturing Director: Tom Debolski

Distributed to the book trade worldwide by Springer-Verlag New York, Inc., 233 Spring Street, 6th Floor, New York, NY 10013. Phone 1-800-SPRINGER, fax 201-348-4505, e-mail orders-ny@springer-sbm.com, or visit http://www.springeronline.com.

For information on translations, please contact Apress directly at 2560 Ninth Street, Suite 219, Berkeley, CA 94710. Phone 510-549-5930, fax 510-549-5939, e-mail info@apress.com, or visit http://www.apress.com.

The source code for this book is available to readers at http://www.apress.com in the Source Code/Download section.

To all the people who have contributed to MythTV in some way, from code to documentation to themes; your finished product is fantastic. Many thanks to our family and friends for making this project possible; without your support such a large project would never have happened.

Contents at a Glance

Contents

About the Authors

STEWART SMITH started writing software after he realized that BASIC on the laptop his dad would bring home from work was a lot more interesting than the word processor. He has a memory of being asked by an installation program for a (now positively ancient) Linux distribution how many i-nodes he wanted on his file system. ("What's an i-node?" was his response.) He now knows more about i-nodes than most, and even his mother has her own Ubuntu system. By day he is a software engineer for MySQL AB working on MySQL Cluster—a high availability clustered database. He's about to hold the record for the most hours spoken by a single speaker at a MySQL Users Conference.

Stewart got his first experience with PVRs after seeing Tridge's TiVo, which he modified to work in Australia. Having heard about this MythTV thing, he decided to buy the cheapest TV capture card he could find to see whether it would work (because at the time, as a student, he didn't want to spend all the money getting a TiVo). It did, and he was hooked. He now wonders how people watch TV without MythTV.

MICHAEL STILL has been working on open source applications for about eight years and has been coding for a lot longer. Michael was originally a TiVo Series1 user (although it had to be hacked to work in Australia), but he switched to MythTV when he moved to the United States a little more than a year ago. He has been playing with image processing and video applications for quite a while now.

Michael is the author of *The Definitive Guide to ImageMagick* (Apress, 2005), which you can find out more about at `http://www.imagemagickbook.com`. He works for Google as a reliability engineer, which is some sort of strange combination of systems administration, cluster administration, and software engineering.

About the Technical Reviewer

JAY R. ASHWORTH is, in no particular order, a programmer, a systems and network admin, a Linux geek, a photographer, a layout artist, a webmaster, and a choral baritone. He has been involved with Linux since 1993 and with MythTV since he built his sister's machine in 2004. His involvement with things written and edited goes back well past RFC 2100 (April 1, 1997) to his earliest days on Usenet in 1983, when he could still read everything in the feed (and the feed took less than a day to arrive . . . at 2400bps).

Mr. Ashworth is 41, lives in Florida, and isn't entirely sure whether he's single. He has blogged at `http://baylink.pitas.com` since about the turn of the century.

Acknowledgments

Thanks to my family and work mates who encouraged me while I pursued this project. Thanks to the MythTV team for a great piece of software, and thanks to the MythTV community for doing such excellent support.

Michael Still

Many thanks go out to family, friends, and work mates who've provided great support over the years. A big two thumbs-up to every contributor to MythTV—you've made it the awesome thing it is today and the even more awesome thing it will be tomorrow.

Stewart Smith

Introduction

The book's website at `http://www.mythtvbook.com` contains various posts about the content of this book and is where you can find the source code to the Google Talk interface discussed in Chapter 13. Where appropriate, we will refer to these posts in the body of the text.

If you have questions about what you see in this book, then the book's website is also the place to go. If you have questions about the content of book, please post a comment at `http://www.mythtvbook.com`. The authors also have personal websites at `http://www.stillhq.com` and `http://www.flamingspork.com`.

This book does not attempt to be a definitive guide or a reference book. MythTV is much too large a project, and much too rapidly moving, for such a book to make sense. Instead, this book focuses on getting you started in the MythTV world and on introducing you to the resources that you need to continue with MythTV projects once you've completed the steps described in this book.

Who This Book Is For

In this book we aim to cover the practical elements of MythTV needed to produce a fully working system. These elements are presented as a series of projects, with one project per chapter. This book is aimed at readers who are interested in MythTV but don't have any experience with it and have limited or no experience with Linux or Unix. We describe all of the required steps to build a working MythTV system.

Although it is possible for you to jump around in the book, we have carefully arranged the chapters to introduce information in a logical order so that dependencies for later projects are ready in time for their use. Where practical, we have provided references to the sections where we discuss these dependencies.

For the best experience of this book, you should follow our recommendations as closely as possible. For example, it is possible to use a different Linux distribution, but at that point much of our advice on package names will need further research on your part. Then again, if you want to experiment, please do; after all, that is one of the core ideas of open source.

How This Book Is Structured

The first chapter of this book discusses what a personal video recorder is and the alternatives to MythTV. Chapter 2 discusses getting ready for the installation, including the right hardware for your needs, the hardware we selected and why, and what file system to use for your video data. We then walk you through installing Ubuntu, which is the Linux distribution you'll use for the rest of the book, and then show you how to check that your hardware is working correctly. Chapter 3 details how to install MythTV and its dependencies and shows how to make sure that MythTV is working correctly. It also covers how to get your remote control working with `lirc`.

Finally, you're ready in Chapter 4 to start recording TV. We discuss issues around TV guide data, such as where to get it from, some basic scheduling options, how to search, and how to resolve recording conflicts. Chapter 5 covers advanced TV recording, including how to detect commercials and how to automatically skip commercials, how autoexpiry of video works, and how to set video playback options. We also discuss transcoding, including how to transcode manually for Sony PlayStation Portable, Apple iPod Video, and software video players and how to automate that transcoding. We show you how to convert commercial detection into cut lists, and how this can be handy, if a little risky, for automated transcoding.

Chapter 6 discusses general MythTV functionality, which you might otherwise miss because it's in the playback menu; this is good to know when working your way around the system. Chapter 7 shows you how to customize the look and feel of MythTV using your own themes.

Chapter 8 works through building remote frontends so you can play more than one recording or video at once or have frontends in different rooms. We show you how to built Macintosh, Linux, and Xbox frontends.

Chapter 9 discusses plug-ins including the weather plug-in, the other-video-formats player plug-in, the Netflix plug-in, the RSS (website aggregation) reader plug-in, the web browser plug-in, the MP3 player plug-in, and several more.

By now you might want to expand your MythTV machine to have more tuners or more disk, which is covered in Chapter 10. You might also want to be able to schedule recordings, delete recordings, and perform other management tasks remotely via the Web, which is all covered in Chapter 11 when we discuss how to install MythWeb.

Chapter 12 covers the DVD playback plug-in and some of the complexities associated with playing DVDs on Linux. We have it all sorted out at the end of the chapter, despite the efforts of the movie studios. We also cover how to copy your DVDs to your hard disk—functionality that is particularly useful for DVDs you watch a lot. We also discuss how to burn your own DVDs.

Chapter 13 discusses our code to display Google Talk and Jabber messages using the MythTV on-screen display. The code also implements an instant messaging control interface for MythTV so you can use an instant messaging client instead of a remote control. We introduce how to use the on-screen display for your own projects as well.

Chapter 14 demonstrates how to get the Internet telephony support in MythTV working and discusses some of the issues of VoIP and SIP. Chapter 15 shows you how to download the latest development snapshots and run the absolute latest versions of MythTV. We discuss when this is a good idea and how to become a more active member of the MythTV community.

Conventions

In all the examples in this book, the paths to menu items are specific to the theme you are using. For this book, we used the MythCenter theme. If you choose to use a different theme, then you may find that you need to use slightly different paths through the menu system to get to the functionality we discuss.

Prerequisites

We discuss software version numbers where relevant in the text of the book. This book covers Ubuntu 6.06 and MythTV 0.20.

Contacting the Authors

You can contact the authors, if needed, via http://www.mythtvbook.com.

## CHAPTER 1

Introducing MythTV

This book is about an open source personal video recorder (PVR) software suite called MythTV. This book presents you with a set of projects we implemented in our own living rooms with MythTV, the theory behind those projects, and the steps needed to make those projects happen. The hope is that this will provide a firm basis for your own MythTV projects, while still being practical enough to be an interesting read.

This book does not aim to be a complete reference to MythTV or a guide for how to develop plug-in modules for MythTV. Although we include a brief overview of the major features of MythTV, we explore only those parts of MythTV that are relevant to the projects in this book, which will include all the parts of MythTV that an average user will be interested in. It will also give you an excellent grounding for further projects with MythTV as well. This book is intended as a hobbyist's project guide, providing suggestions about what sort of projects you could take on and how we went about implementing our own versions of those projects.

Instead of including exhaustive coverage of features that few people use, we'll provide pointers to how to find out about those features, and we'll cover the 80 percent of MythTV's functionality that everyone will find useful. That way, this book will be genuinely useful to people setting up MythTV, instead of a boring reference manual.

We discuss the projects we undertook using the components we selected. Where appropriate, we explain the alternatives available at the time of writing, why we chose the paths we did, whether we think that in hindsight these were the right decisions to make, and how to complete the projects using the same decisions we made. You are free to choose different components for your MythTV system, but it is impractical for us to document every possible combination of every possible component. We do provide as many pointers as possible to further information as we discuss particular points, and it is our intention to add future information to the book's website when appropriate.

Defining What Personal Video Recorders Are

This book discusses a PVR suite called MythTV. A PVR—also sometimes referred to as a digital video recorder (DVR)—is a computer system that allows you to easily specify which TV shows you want recorded and then makes sure the recording occurs without further human intervention. These computer systems are often referred to as *home-theater PCs*, although there doesn't need to be any inherent difference between a regular personal computer and one of these systems, as you will see in Chapters 2 and 3 when we show how to build one from scratch. The underlying concept is, of course, that you never need to worry about recording the shows you like to watch—you can just come home from a hard day at the office or at school, and the computer will have a bunch of shows ready for you to watch.

This concept is often referred to as *time shifting* because you're shifting the time at which you watch the shows. A good PVR suite quickly becomes more than just that, though. Once you have a system to record all the shows you want recorded, then why not use that system to ensure that your favorite podcasts are available when you want them? Or to provide access to your music collection, online photo albums, and so forth? A PVR project can quickly become a hub for all the entertainment in your home, and although you might not expect that outcome when you start your own MythTV implementation, ending up with a full-convergence system isn't bad.

The projects in this book will cover how to construct a PVR using MythTV and then explore some of the other cool projects you can implement to turn that MythTV PVR into the hub of your living-room entertainment. Many other projects are possible as well, but the ones in this book will give you a good grounding.

MythTV isn't the only way to get a PVR into your living room, although it is one of the most flexible alternatives. We discuss other options in the "What PVR Systems Are Available?" section.

It should be fairly obvious from this short definition why people want PVR technology—it takes a bunch of the manual steps out of your entertainment lifestyle and therefore leaves you with more time for other activities. It also reduces the odds that you'll miss some TV that you just must see.

Understanding the Components of a PVR System

Now that we've convinced you that you want a PVR system, we'll describe the components of such a system. These components are standard between the various vendors for PVR solutions, although the exact implementation will vary. Figure 1-1 shows these elements.

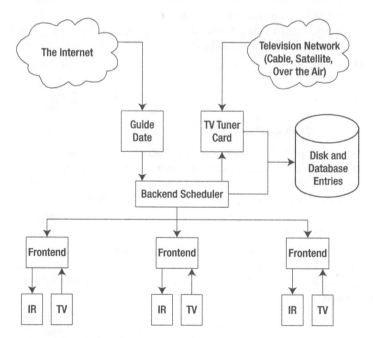

Figure 1-1. *Components of a PVR system*

You can see from Figure 1-1 that a PVR system has a number of independent functional units that all interact to produce the final result. On the TV-capture side of the equation, a guide data system downloads the list of which networks are showing which shows and at which time from somewhere on the Internet (sometimes supplemented by a guide broadcast over the air). This system then populates the program database, which the scheduler uses to decide which shows to record and at which time. The backend then instructs the tuner card to tune to the right station to receive a given show and then saves the video and audio data from the tuner into a file on disk. Also, the backend maintains database entries so it knows what shows you like, when to record them, and what shows it has already recorded.

One advantage of having a centrally scheduled system is that the scheduler can determine the best order for shows to be recorded. For example, if you have only one tuner and there are two shows that should be recorded at the same time, then it can look at its guide data for the future, determine whether one of those shows is repeated later, and therefore resolve the conflict by scheduling that show to be recorded on its repeat airing. The scheduler also knows which shows form a series, and it is trivial from the PVR user interface to ask for an entire series to be recorded. Finally, if you have more than one tuner, then the scheduler can take full advantage of that to capture more than one show at once and therefore have more shows available for you to watch at any given time. A good PVR backend will also ensure that shows you've already seen aren't rerecorded unless you want them to be.

The other half of a PVR system is the frontend. This provides the user interface, including helping you look through program guides to determine what to record, displaying videos to you once they have been recorded, accepting infrared (IR) commands from the remote control to control that playback, and so forth. Many MythTV users utilize the computer monitor as a playback device, which is a perfectly acceptable solution in many cases, especially with many LCD and plasma TVs now having VGA or DVI inputs. Chapter 2 will show how to use the TV card to send the playback to a standard TV instead if that suits your needs better.

Some PVR systems' frontends also do additional processing, such as changing the aspect ratio of the video being played back, pausing live TV as well as recorded TV, and automatically skipping through advertising. All of these features are available in MythTV, and we will discuss them as we come upon them.

What PVR Systems Are Available?

A PVR system needs several hardware components to be practical. You need a reasonably fast CPU (or embedded video hardware, although that is an approach more common on commercial "appliance" PVRs), plenty of storage, and video input and output hardware. All of these components were outside the price range acceptable to most home users until the 1990s. When the hardware became available to build PVR systems, the number of PVR systems available exploded as hobbyists around the world saw the possibilities available to them.

The following sections don't intend to be an exhaustive coverage of all these systems, but we will give you a brief overview of the most popular competitors in this space so you can make an informed decision. It might be that for some reason MythTV isn't the right solution for you and you're better off with an alternative.

Home-Grown PVRs

Some of first entrants into the PVR space were hobbyists who wanted to be able to record shows when they weren't home and wanted to be able to easily fast-forward through the advertising that is so common on TV these days. These "home-grown" PVRs are built on top of standard PC hardware, although all of them are available in prebuilt forms as well.

MythTV

MythTV (http://www.mythtv.org/) is the PVR system that is the focus of this book; all of the projects described are directly applicable to MythTV, and although other PVR systems might have equivalent functionality, the instructions in this book won't apply to them. MythTV is open source, released under the GNU General Public License, which means the source code that the programmers wrote to implement MythTV is freely available and that you as a user have the right to modify that code to suit your own needs (although you are asked to release your changes to the community and are actually required to release those changes if you distribute your modified version).

The big advantage of open source software for projects like the ones in this book is that extensibility is the killer application for your living room—having a system that is tightly integrated into your various digital entertainment systems and that is customized for your own use is very attractive to a large number of people. The ability to build your own system is also appealing to hobbyists, because it matches their own intentions.

We won't say anymore about MythTV here because we cover it much more extensively throughout the book.

SageTV

SageTV (http://www.sagetv.com/) is one of several applications that you can run on your existing Windows or Linux PC to turn it into a PVR. SageTV is commercial software that, at the time of this writing, costs just less than $100. One advantage of packaged systems such as SageTV is that you get an application that is ready to go as soon as you install it. In fact, SageTV also sells hardware that is certified to work with its system. On the other hand, prepackaged systems are less extensible than open solutions and are quite expensive compared to some of the dedicated hardware or free PVR options.

Windows Media Center

Microsoft is a relatively late entrant into the PVR market with its "Media Center Editions" of Windows XP (http://www.microsoft.com/windowsxp/mediacenter/default.mspx) and the just-released Windows Vista (http://www.microsoft.com/windowsvista/features/forhome/mediacenter.mspx). If you want a solution that is based on Windows, then you might want to consider a Windows Media Center installation. You can find one person's side-by-side comparison of MythTV and Windows Media Center here: http://bruceshankle.blogspot.com/2006/04/windows-media-center-vs-mythtv-or.html.

One advantage of a Microsoft Windows–based solution if you're already using Microsoft Windows is that you can continue to run the other Windows software that you are used to using. However, it's probably unlikely you'll want to use the computer in your living room as a desktop machine, because it's going to be plugged into a TV, and that means other people won't be able to use your living room for entertainment. Then again, if you're building your

PVR for another room such as a bedroom or you are space-, power-, or cash-constrained, then this might be an attractive alternative.

Additionally, at the time of this writing, Microsoft doesn't sell the Media Center version of Windows directly to the public; you have to buy a new machine with it preinstalled (http://www.microsoft.com/windowsxp/mediacenter/howtobuy/default.mspx). If it's not preinstalled, this means you can't use an existing computer for your PVR project. Additionally, the hardware you can buy preinstalled with Windows Media Center Edition is generally more expensive than regular PC hardware, and you don't have as many options. An example is the popular Mini-ITX system that many people use because the system's near silence makes it a perfect choice for the living room—many retailers don't stock these systems preinstalled with Windows Media Center.

Yahoo! Go (formerly Meedio)

Yahoo! acquired the Meedio PVR product in April 2006 and has released it as a free PVR application for Windows computers (http://mobile.yahoo.com/go/). Although we haven't used Yahoo! Go, we suspect this is a better option if you're after a simple-to-implement PVR system and were otherwise considering SageTV or Microsoft Windows Media Center Edition. This is especially true because Yahoo! Go is free, so if you do want to upgrade to something else later, it's not as painful. Remember, though, that Yahoo! Go is closed source, so you won't be able to extend it in the ways we demonstrate in this book.

Embedded/Appliance PVRs

It's possible that building your own PVR system isn't attractive to you for some reason. A number of commercial PVR systems are now available for purchase. These systems are generally not as configurable or as extensible as the hobbyist systems, but they are simple to implement and often initially cheaper than building your own system (especially if you consider the amount of time it takes to set up your own PVR system).

TiVo

TiVo (http://www.tivo.com/) was the first manufacturer of a commercial PVR system. TiVo looks similar to a DVD player, and in fact some models have a DVD writer built in (see Figure 1-2). Because it is a commercial system, As long as you do what TiVo expects you to do with the device, then it will "just work." Specifically, the following requirements apply to TiVos:

- You have to be using your TiVo box in the United States to get full TiVo functionality. TiVo by default can download TV guide information only from the TiVo site, which supports only the United States at this time. TiVo has attempted to break into the U.K. and Japanese markets but no longer sells devices in those places. If you're outside of the United States, you can use your TiVo box if you have NTSC video available, but you'll have to set everything up as manual recordings.

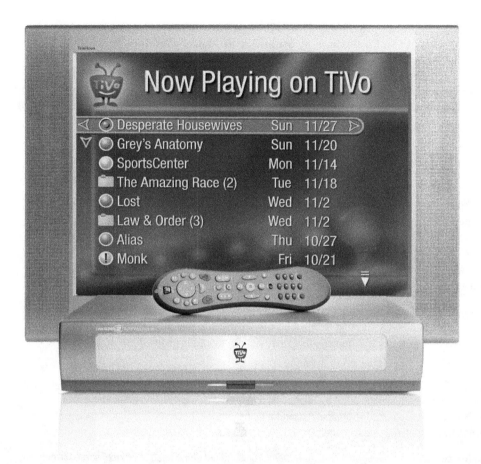

Figure 1-2. *TiVo's list of recorded shows known as a "Now Playing" screen. © 2006 TiVo Inc. All Rights Reserved.*

- For the longest time your video input had to be either SVideo or U.S. analog cable. Digital cable tuners were not available, and if you live in a PAL or SECAM country, then you're out of luck. TiVo finally released a digital cable–capable TiVo Series3 in mid-2006. See the "NTSC, PAL, and SECAM" sidebar for more information about NTSC, PAL, and SECAM.

- You have to be happy with the size of the hard disk on the device. TiVo does not provide hard disk upgrade options, although you can buy a whole new TiVo with a bigger disk. It is possible to perform after-market upgrades, but they will void your TiVo warranty.

- You have to pay for an ongoing service contract to get guide data, and TiVo will upgrade your device to the latest version of the operating system remotely. There have been examples of upgrades that have arguably reduced the amount of functionality available on the device and upset users. For example, TiVo has recently been working on displaying advertising on TiVo machines (http://news.com.com/2100-1041-5178017.html) and limiting the length of time you can keep a recording (http://www.pvrblog.com/pvr/2005/09/tivo_72_os_adds.html).

If you're willing to modify your TiVo box, then these requirements aren't as much of a problem. One of us, for example, was quite happily using TiVo in Australia for quite some time. This TiVo needed a specially tweaked operating system installation to be able to download Australian guide data, a different TV tuner installed (which involved soldering, and so forth), and didn't receive support from TiVo because of the modifications needed. For more information about getting TiVo working in Australia, check out the Australian TiVo community site at `http://minnie.tuhs.org/TiVo/`.

NTSC, PAL, AND SECAM

The pictures displayed by your analog TV are encoded when transmitted on the cabling used by your various entertainment devices. These encodings are also often used to distinguish the actual media such as video cassettes and DVDs as well, although this is not a strictly accurate description of what is occurring. Refer to the "PAL" Wikipedia page listed in this sidebar for more information about this.

Unfortunately, three competing encodings exist, which makes it much harder to use audiovisual equipment and media from other countries. The three competing standards are as follows:

- NTSC (which stands for National Television Systems Committee) is the encoding used in North America, Japan, and Korea. Starting out as a black-and-white format in 1941, it progressed to including color support in 1953. You can find more details about the history of NTSC at `http://en.wikipedia.org/wiki/Ntsc`.

- PAL (which stands for Phase Alternation Line) is the encoding used for many parts of the world. The PAL encoding was introduced in 1967 by Telefunken and is based on NTSC. PAL offers better video quality than NTSC, which was the original reason for its development. You can read more technical details about PAL on the Wikipedia page at `http://en.wikipedia.org/wiki/PAL`.

- SECAM (which stands for *séquentiel couleur à mémoire*) is the encoding used in France, some former communist countries, portions of Africa, and parts of the Middle East. Although development started in 1956, it was not released until 1967, and SECAM TVs were very expensive when released. Many European countries at one time used SECAM, but most have now converted to PAL. Wikipedia's page on SECAM at `http://en.wikipedia.org/wiki/SECAM` contains lots more details.

(For purists, these are the *color* signal encodings, and there are often further difficulties involved with the encoding and scan-line count of the video signal. However, these are, for the most part, issues you don't have to deal with—your tuner card will deal with them for you.)

Cable-Company DVR

Your cable or direct broadcast satellite (DBS) satellite company probably also provides a PVR option. It's hard to comment on these generically, because they vary a lot. We have found with the several we have used that they are very expensive, have limited storage, or delete shows after a short period of time, even if you haven't watched the show. Your shows will also become inaccessible if you cancel (or fail to pay the bill for) your cable service, and it's usually impossible to get the programs off the unit except by hooking it up to a VCR. Be careful of these factors if you are considering using a PVR provided by your cable company. If you already own one of these and want a MythTV machine as well, then you'll need to look into Apple FireWire access for your model of set-top box or look into some combination of video cabling and IR blasting.

Why Use MythTV?

This book discusses MythTV, mainly because it's an open source package that is therefore extensible and well suited to the projects covered in this book. It's also free, which makes it attractive, although you should always factor in the cost of the hardware and your time to implement the system when comparing it as an option to a commercial solution.

MythTV also integrates with the DVD player in both your living-room PC and your Microsoft Xbox, if you have one. It plays your MP3 collection and videos from sources other than TV (for example, AVI files and videos from personal video cameras), displays photos from your photo collection, gives you access to weather updates for your ZIP code, and displays news updates using RSS syndication technology. MythTV will even help you manage your movie queue with Netflix.

MythTV also has a versatile plug-in system, which makes it easy for developers to implement their own functionality, and has a relatively simple XML-based system for defining themes if you want a custom user interface. Then again, the default theme is pretty nice. Figure 1-3 shows what the front screen of a custom MythTV installation looks like.

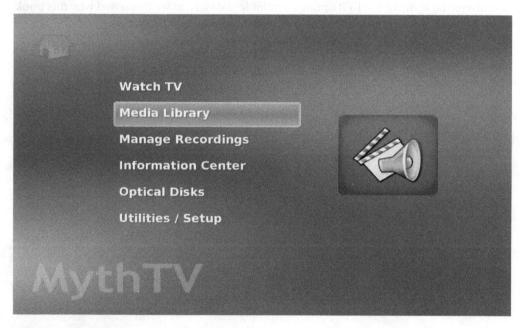

Figure 1-3. *The MythTV welcome screen*

MythTV runs on standard PC hardware. The only piece of slightly nonstandard hardware you need is a TV capture card, and these are quite common now. Chapter 2 discusses how to select the hardware for your MythTV installation and walks you through the process of setting up the system.

ABOUT OPEN SOURCE VERSION NUMBERS

Throughout this book the version numbers for software products might appear very low if you've come from a Microsoft Windows–based world. For example, MythTV is only at version 0.20 at the time of this writing. However, this is because the number style for many open source applications starts at zero. For example, 0.20 commonly means that this is the 20th public release of an open source application, which means it's quite mature.

Generally, the first number will be incremented when a major change affects compatibility or, if from 0 to 1, when the author thinks the program is both feature complete and sufficiently well tested to be considered a 1.0 release—some prominent and useful open source programs have never reached version 1.0 (and some, like the MAME game emulator, explicitly never will).

Conclusion

This chapter introduced the concept of a PVR, gave an overview of how a PVR system works, and then discussed some of the PVR options available today. Next we discussed why this book focuses on MythTV. You're now ready to move on and build a MythTV system.

CHAPTER 2

■ ■ ■

Getting Ready for the Install

In this chapter, we cover the basics of a MythTV installation—from selecting hardware (a TV card, disks, and a CPU) to selecting some operating system configuration options (such as which file system to use) and installing the operating system. We'll use Ubuntu Linux throughout this book, although MythTV should run on any recent Linux distribution. We selected Ubuntu because of its ease of installation, excellent packaging, and nice user interface consistency. We've also chosen Ubuntu 6.06 LTS instead of the more recent 6.10 version because it is a "long-term support" release and will be supported with security updates through 2011.

Ultimately everything is your own choice. Some choices, though, will make the process easier or harder on you. We've done a lot of research for you and present our opinions here (as well as tell you what we choose to do). This should help you make the decisions you need in order to get the best MythTV setup.

After finishing this chapter, you will have working hardware ready for installing MythTV, which is covered in Chapter 3. You'll need to follow the steps in this chapter and Chapter 3 before any of the rest of the book will be much help to you.

Selecting Your Hardware

One of the more important decisions to make when implementing your MythTV system is what hardware to use. The hardware is certainly the most expensive decision you'll have to make and the hardest to change if you decide you made an error. The following sections therefore discuss the various pieces of hardware you'll need to arrange in order to implement MythTV as well as the factors to consider when purchasing. We'll also discuss the hardware we used when developing our own MythTV systems, which is the hardware we use for all the examples in this book.

We'll also give special attention to areas where you can easily expand your system later. This is so you can start small and grow later—a well-advised strategy. It also means you can take advantage of increasingly affordable hardware.

Starting Small

Starting small is a great idea because it lets you test the software first and then dedicate hardware (and the associated money that hardware costs) to a MythTV setup. Stewart started really small—he bought the cheapest TV capture card he could find and added it to his existing desktop PC (an ever-so-loud Pentium 4 that you would never want in the living room).

Because Stewart already used Linux as a desktop operating system, it wasn't a problem to just install MythTV and run the frontend portion only when needed. (The part of MythTV that does the recording—the backend—runs in the background, allowing you to run other applications while shows are being recorded.)

Michael's experience was a little different. After using a TiVo in Australia, he knew he couldn't live without some sort of PVR technology in his house. Michael bought a new machine specifically for use as a MythTV machine, although he also used it for other tasks such as serving email and sharing files.

However, not everybody runs Linux as a desktop operating system, and not everybody is willing to spend too much time trying to get a software package installed and configured correctly.

If you want to try MythTV before deciding to build a specific machine for it, then you should look at KnoppMyth, which is a Linux LiveCD with MythTV built in. (A LiveCD is a CD you can boot your computer from instead of the hard disk, which allows you to run a different operating system than what is installed on the machine's hard disk without affecting anything installed on your machine.) You can find out more about KnoppMyth at `http://www.mysettopbox.tv/knoppmyth.html`.

MythTV's Architecture

MythTV has two main components: the backend and the frontend. The *backend* is responsible for capturing video, encoding it, and writing it to disk. The *frontend* displays a user interface on a TV or computer monitor and allows you to play recordings. The frontend also has several plug-ins available, which perform tasks as various as providing current affairs, web surfing, arcade game console, MP3 playback, and more. Chapter 9 covers most of those plug-ins. Typical MythTV setups include running both the frontend and the backend on the same machine or having the frontend separate from the backend. It's also possible to have multiple frontends and multiple backends. Chapter 8 covers running with more than one frontend.

How fast your CPU needs to be depends on a number of factors. Recording digital TV in MPEG-2 takes up a trivial amount of CPU time, and any CPU you can buy today will happily accommodate several streams. In fact, many video capture cards will do the video compression in hardware. Thus, if the machine you are specifying is going to run the backend, you need to be aware of how many video capture sources you intend to have, how many will be recording at any given time, and whether those sources implement video compression themselves.

Encoding analog video as it is recording (especially from lower-quality signals) can take up more CPU power—even 60 percent of one of the author's AMD Sempron 2600+ CPUs. Digital capture uses a few percent at most. Digital playback can use some CPU, though—maybe 10 percent for standard definition. However, high definition requires much more processing power—typically 60 percent to 70 percent of a Sempron 2600+. This is mainly because the high-definition video frames are each much larger than standard-definition frames. The CPU usage for playback will be lower if you use an X video (xv)–compatible video card.

Finally, you need to decide how much post-processing such as commercial detection and transcoding you are going to do on the machine. It is possible to spread this workload across multiple backends, but the most common setup is to have just one backend running with one or more frontends.

A general recommendation is to buy the fastest CPU you can afford, within the constraints of the case and motherboard you have selected. Stewart is currently using an AMD Sempron 2400, and Michael is currently using a 3GHz Intel Pentium 4.

If your MythTV box will be sitting in the living room, you don't want it producing much noise at all. Stewart uses a Shuttle "bare-bones" system (the motherboard is integrated with the case). It's small, it's quiet because it has a special CPU heatsink that uses a very quiet fan, and it has all the ports you'd expect on a computer. In regular operation, the loudest part is the hard disk, and you really have to struggle to hear it. Other manufacturers also have quiet cases—some even designed for media applications. Check out the options and remember that if you want more capture cards than you can fit in a small box, you can always add other backends that can be located in rooms where noise isn't so much of a problem or use USB capture hardware.

Graphics Card

In reality there are really only three manufacturers of chips that drive graphics cards: Intel, ATI, and NVIDIA. Since 3D performance doesn't matter for displaying TV, you are not going to need a whizz-bang gamer-style video card for your MythTV box. What's more important is the ability of the card to handle video overlay reliably and do NTSC/PAL video output. Selection can depend on if you are going to be using a DVI/VGA signal to your TV (or computer monitor pretending to be a TV) or an S-Video or composite connection.

MONITOR INTERCONNECTS

Video Graphics Array (VGA) and Digital Visual Interface (DVI) are both names of monitor interconnects. That is, they are transmission formats for data from your computer to its monitor. VGA is the older of the two, with the original implementation dating back to 1987. A VGA connector has 15 pins in a D shape, and data is transmitted with an analog encoding. At that time, VGA was also used to refer to the 640-pixel-wide and 480-pixel-high graphical display it was used to implement. In this book we are always referring to the monitor interconnect when we use the term VGA.

DVI, on the other hand, has an up-to-29-pin plug, which is fairly easy to recognize because it is wider and has a big flat pin. It's a digital transmission format released in 1999, and its digital nature produces a crisper, more defined picture on digital monitors. Whether this is important for you will depend on the quality of your display (be it a projector or a monitor) and the quality of the TV signal you are capturing.

If you want to know more about VGA, you should read the excellent Wikipedia page at http://en.wikipedia.org/wiki/Vga; for more about the DVI format, try the excellent Wikipedia page at http://en.wikipedia.org/wiki/Dvi.

Intel is generally rather friendly to the free software community, and its graphics cards are usually well supported. If you can get a motherboard with an onboard Intel VGA or DVI implementation that supports the output you need to get to your TV, these are probably safe bets.

NVIDIA and ATI are another story. They have proprietary, binary-only drivers that they ship for Linux. This does not help the free and open source community at all, and any problems you have while running them will largely be ignored by developers (because they cannot debug the problem with binary-only drivers loaded). There are also free, open source drivers for some ATI and NVIDIA cards—and since the open source drivers have little trouble with 2D and video overlay, they are quite suitable for MythTV work. We recommend you give the open drivers a try as your first option and fall back to the closed drivers only if you absolutely must.

For NVIDIA cards, there is the free nv driver and the closed source nvidia driver. The free program nvtv can enable TV-out through composite/S-Video for some NVIDIA cards. The proprietary NVIDIA driver will work with more cards and support 3D acceleration as well. The 3D support is not necessary for MythTV; but if you use your frontend machine for other purposes, this might be more important to you. For the typical appliance-type MythTV frontend, it's not an issue. Additionally, it is quite easy to install, and with Ubuntu, it's quite painless.

The binary-only ATI drivers have historically been harder to install and configure than the NVIDIA ones. With Ubuntu 6.06, this is largely no longer the case. However, many people prefer NVIDIA cards because of the historically easier-to-use setup and even more up-to-date and stable proprietary drivers.

The good news is that since you don't need a lot of graphics capability, the older cards might be OK for your setup. However, this is not always true at higher resolutions. If you use a large screen (Michael uses a 24-inch Dell LCD monitor as a TV), you need to have a graphics card that has enough video memory to use the full resolution. Some onboard Intel cards might not—so check first! You also need drivers that support xv—an acceleration architecture for displaying video in the X Window System, the system that puts pictures on the screen. Some ATI cards don't support xv, and the video will be choppy. Additionally, if your video card doesn't support xv, then more CPU will be used during playback because the frontend will need to emulate xv with software. Both authors have now settled for NVIDIA cards in their MythTV systems. Resolution and xv support are particularly important if you intend to play back high-definition TV.

Disk

Like everybody who ever has used (or indeed will use) a computer, the disk is something you will never have enough of, and it will never be as fast as you want. Storing video has traditionally been a rather demanding application, but this is not so true these days because of good compression algorithms and fast CPUs. Our main concern is disk space, not speed (external USB disks provide more than enough speed).

How Much Disk Space Do You Need?

This is an interesting question. It boils down to three factors:

- At what quality will you be recording?

- For how long do you want to keep recordings?

- Are you going to store other data (such as music, photos, or other videos) on the MythTV machine?

When doing the calculations for how much disk space to buy, remember to set aside about 10GB for the operating system. For example, a 250GB disk will get you about 234GB of formatted capacity, minus 10GB for the operating system, which leaves you with 224GB of space for recordings. It's traditional to have the operating system installed on a separate partition on the machine's disks, and you'll possibly have separate partitions for /var and swap as well. These separate partitions are generally used to minimize the impact of having a partition fill up. For example, if you create many large log files in /var, having that separate will stop the rest of the machine from being adversely affected. Additionally, it gives you the flexibility to use logical disks (which we mention later) and different file system types on different partitions.

Recording Quality

Stewart started off thinking "as long as it's about as good as Long Play on VHS, I'm happy" when he first built a MythTV machine. However, if he'd just bought a high-definition projector, some really nice sound gear, and an armchair with a fridge to keep the beer cool, this would not have been acceptable quality!

You might source your TV from an analog signal through a cheap TV capture card. A good analog signal can be compressed well and give you a good-quality image during playback. If you compress it more, you might still be happy with it and pleased with the disk space that you save, but the image might not be as good. Stewart has found that compression settings on a well-received analog signal that approximates a Standard Play recording on a good VCR uses about 1GB per hour of TV. A DVD is noticeably better quality than the recording; with a digital TV card, the quality is as good as DVD. With poor reception and analog, though, it can be as much as 2GB per hour of TV.

If you're sourcing your TV from a digital TV card or any of the analog cable TV cards that supports MPEG compression in hardware, then you will be receiving raw MPEG-2 streams of video. Standard-definition TV is usually 2GB to 3GB per hour. A high-definition program can use up to 7GB per hour of TV. You can record these streams without recompressing (transcoding) it, and it will look *exactly* like you were watching live TV through the digital tuner. However, you might choose some further compression to save disk space. You can also configure MythTV to recompress only certain programs. (For example, you might want your recordings of *Law and Order* to be in high definition, but you'd rather save space on *Wheel of Fortune*.) We discuss this in Chapter 5.

It's important to note that the larger the gigabytes-per-hour rate is, the more disk bandwidth is required. This will probably be an issue only if you consider recording multiple streams at once while playing back programs.

How Long Will You Keep Recordings?

This can depend on your watching habits. Stewart has found that with the shows he records (and deleting some shows after he has watched them) with his 250GB disk, he keeps recordings for about a month before they autoexpire to make room for new ones. Michael, on the other hand, currently has 700GB of storage and has found that to be "about right" for him.

You can heavily customize which recordings autoexpire first (if at all) and make programs never autoexpire. With the new 0.20 release of MythTV, Live TV recordings are now treated like regular recordings and have their own autoexpire settings. You can also at the touch of a button convert the program you were watching into a recording. Refer to Chapters 5 and 6 for more information.

Where Will You Store Other Data (Music, Photos, Videos) on the MythTV Machine?

Consider whether you are going to store other multimedia on the same set of disks in the MythTV machine. Here are some numbers to think about:

- Your CD collection (100MB per CD for good quality).

- Your digital photos (largely depends on your camera and how often you take photos). This could be a few hundred megabytes or many, many gigabytes.

- Other videos. For example, think about the videos you download from Google Video, YouTube, and so on. These vary in size (and quality), and you might not want to keep them for long.

Choosing a Disk to Buy

You might also choose to buy more than one disk and set them up in a RAID or logical volume manager (LVM) configuration—see in the "Disk Layout, Logical Disks, and Meta Disks" section later in this chapter for more detail on those. If placing the disks in the living room, you'll want them to be relatively quiet. Usually the manufacturer lists how much noise the drive makes in various modes of operation on the specifications page on its website. However, even the "noisy" drives can be rather quiet when in a computer case inside a cabinet with the TV on, and many disks also have options in the BIOS for speed versus noise trade-offs. Do not consider 5400rpm disks; you want the extra speed that 7200rpm gives you. Additionally, the slower disks are getting increasingly hard to find and are not any cheaper. At time of this writing, the extra cost for 10,000rpm disks isn't worth it—instead, it is more affordable to get two 7200rpm drives and configure them in a striped RAID setup if you're really worried about speed.

We previously talked about how much disk space recordings might use. The general recommendation for buying a disk is to buy the biggest you can reasonably afford. There is often a sweet spot that gets you the most gigabytes per dollar. When Stewart was setting up his MythTV box, this was 250GB drives. The sweet spot is now at around 500GB drives, and by the time you read this, it will be probably higher again.

Should you buy Parallel ATA, Serial ATA, USB 2, or some new fancy connector? You are going to need an internal disk for your MythTV box, which means Parallel ATA (PATA) or Serial ATA (SATA). Stewart chose PATA because it was cheaper and he didn't see performance as critical for him. However, the price difference between PATA and SATA is now trivial (if there is one at all), and SATA is certainly the way of the future. Michael, on the other hand, has decided to store most of his data on external USB 2 drives (he has one internal SATA disk and six external USB disks). He is also thinking about iSCSI or ATA-over-Ethernet as a possible network solution so that the disks can sit in another room from the MythTV box. He has also been experimenting with consumer network-attached storage (NAS) devices to this end. It is certainly advisable to start with a simple setup and look at expanding in the future. This leaves you with much fewer things to go wrong during your initial setup.

CD/DVD Reader/Writer

These days DVD writers are very cheap. There is little point in just buying a DVD reader and next to no point at all in buying just a CD-ROM drive—if you can even find one. The only debate you're likely to have is whether you'll get a single- or dual-layer DVD writer. It all depends what you're going to want to write to DVD and the quality at which you recorded it.

Having a DVD writer in your MythTV system makes it much more like a VCR. With a VCR you can tape something from the TV and label the tape for later playback. A DVD writer lets you do the same. It makes your MythTV box similar to the DVD recorders you can buy at the local electronics store. You can read more about how to set up DVD burning with your MythTV machine in Chapter 12.

Be careful, though; depending on where you live, you might not have the legal right to record programs from the TV and watch them later—even with a conventional VCR (which have been commercially available for the home market for more than 30 years). Also, you might be allowed to view the recorded programs only once and not use the Rewind button. Check local laws if you're concerned.

Video Capture Cards

Selecting a video capture card is probably the most complicated part of building your MythTV system, mainly because you have lots of choices. Choosing the right one can look daunting. The first decision you'll have to make is deciding whether you want an analog or digital card. Because many places in the world (including the United States and Australia) plan to phase out analog broadcasts between 2006 and 2010—Holland did so at the end of 2006—digital is a safer bet. These dates are not set in stone, though, so you might find that your home country has no announced plans to turn off analog transmission or that the government has backed away from previous deadlines. The motivation for turning off the analog transmissions is clear, however; it will free up large amounts of the available broadcast spectrum for other uses. On the other hand, analog cards can be quite cheap and can require less processing power and disk space. Analog cards can also be a more sensible option with some cable TV services (for many parts of the United States this means not requiring a set-top box and setting up an IR blaster).

If an analog card is being sold locally, it should support the color encoding (either PAL, NTSC, or SECAM) of your region. Most cards can easily switch between several of these with a simple software configuration change.

There are also differences in digital transmissions. There is the simple difference of standard definition versus high definition. If your TV isn't high definition, there is little purpose in bothering to capture the high-definition signal, and you can probably safely ignore high definition (unless you plan to upgrade your TV shortly).

Finally, remember to check that your planned card is supported by Linux and by MythTV. This can be quite a subtle choice—for example, not all the Hauppauge cards are supported by Linux, but the ones that are supported work very well and are probably the most common analog cable card in use by MythTV users in the United States. You can find a matrix of the cards supported by MythTV at http://www.mythtv.org/wiki/index.php/Video_capture_card. Table 2-1 summarizes the relevant portions of that matrix at the time of this writing.

Table 2-1. *Recommended Video Capture Cards*

Linux Driver	Example Cards	Hardware Compression?	Does Driver Ship with Ubuntu?
bttv	Aimslab Video Highway Extreme	No	Yes
	ATI TV-Wonder		
	ATI TV-Wonder VE		
	AVerMedia DVB-T 771		
	Hauppauge WinTV-Go		
	Hauppauge WinTV 28061 (revision B226)		
cx88-dvb	Hauppauge Nova-T	No	Yes
	Hauppauge Nova-S	No	Yes
dvb	Technisat AirStar HD-5000	No	Yes
FireWire	Some set-top boxes	No	Not applicable
IVTV	AVerMedia M179	Yes	No
	Hauppauge PVR-150		
	Hauppauge PVR-250		

continued

Table 2-1. *(continued)*

Linux Driver	Example Cards	Hardware Compression?	Does Driver Ship with Ubuntu?
	Hauppauge PVR-350		
	Hauppauge PVR-500		
saa7134	Compro VideoMate TV Gold Plus	No	Yes
Linux 2.6.15+	DVICO FusionHDTV5 RT Gold	Yes	Yes
Linux 2.6.17+	DVICO FusionHDTV5 Gold	Yes	Yes

You can find a much more detailed description for each card available at `http://www.mythtv.org/wiki/index.php/Video_capture_card`. However, some recommendations can make your life a lot easier, so let's explore some recommendations for video capture cards.

Selecting an Analog Card

If you decide to select an analog tuner card, then your next step is to decide what TV sources you want to record. If you plan on recording TV that is broadcast free over the airwaves (what we refer to as *free-to-air TV*), then the Brooktree 878 (BT878) series is well supported, cheap, and easy to find. These cards often have an aerial input and an S-Video input, which means they can record these free-to-air sources as well as video from a set-top box such as those deployed by Foxtel in Australia and the Dish Network in the United States. To find other well-supported cards, the best bet is to look for cards in stores that are listed on a "Supported Hardware" list. The Video4Linux (V4L—the system that supports analog video capture cards on Linux) wiki maintains a list of supported cards at `http://linuxtv.org/v4lwiki/`.

The wiki also has specific troubleshooting information for each driver. If you find out new information, you can also add it to the wiki so others can learn too.

For these free-to-air sources, we will focus on the BT878 cards because they are the most common, and those are the ones for which we provide installation instructions in the following sections.

If your TV is coming from analog cable, then you will need a video capture card with an analog cable input. This will be the case if you use an American cable network such as Comcast, which just provides you with a coax plug on the wall. The advantage of these cable providers is that you don't need to use IR blasting, because the tuner on the video capture card can do the right thing. Your best choice for these analog cable providers are the Hauppauge PVR cards, although not all of them are supported by Linux, so check before you buy one. The PVR 150, PVR 250, PVR 350, and PVR 500 are all excellent choices at the time of this writing. These cards use the IVTV driver, which is also discussed later in this chapter. The only item that requires caution in selecting a Hauppauge tuner card is that the same model of card ships with different tuner modules depending on when it was manufactured, and it sometimes takes a while for the IVTV drivers to support newer tuners.

Selecting a Digital Card

If your digital source is free to air, then there is again an excellent wiki that can help you choose what card to buy because it lists supported hardware. The Digital Video Broadcast (DVB) wiki, at `http://www.linuxtv.org/wiki/`, has this list of supported hardware.

More commonly now, many small (or large) online stores will list whether the devices they are selling work with Linux. This is indeed the case for `http://www.digitalnow.com.au/`, which sells DVB-T cards in Australia. This shop is often recommended on user group mailing lists to those looking for well-supported digital capture cards in Australia.

Other Issues to Consider

Some cards support two or more tuners. These are great if you ever want to record one show after another or two shows at once. They are probably worth the extra expense because they are often cheaper than buying two cards. Adding USB capture cards can also be an option because many of them are well supported. However, beware of hardware revisions—sometimes the manufacturers change the hardware revision, and the Linux drivers no longer work or require tweaking to get working. One of your authors has been bitten by this, but luckily he had the knowledge required to modify the driver. You can read more about adding tuners to your MythTV machine in Chapter 11.

Now that we've discussed a lot of the hardware choices you have to make, what choices have we made? Often some of the best advice is from people who say, "This worked for me." So let's discuss what we ended up using.

Stewart's Hardware Choices

I chose a Shuttle SN85G40, which was a current model at the time. It supports 32- and 64-bit AMD processors, giving me an upgrade path. I chose an affordable processor (the AMD Sempron 2600+), and with a cheap analog capture card and bad reception, encoding video was using a lot of CPU, and I was almost wishing for a faster one. I went the cheap option and bought only 512MB of RAM (as a single stick and have since added a 1GB stick). With the benefit of hindsight, it might have been better to buy 1GB at the start because I found that when I was watching, recording, and transcoding, some memory pressure occasionally caused recordings to drop frames.

I bought the sweet spot of 250GB of disk. I bought a 7200rpm Western Digital drive. I've used WD drives for several years now and have not had any problems. You will always find somebody with the opposite opinion. This is enough disk for my TV recordings. I don't tend to keep recordings stored on the MythTV box. If I did, I'd want a lot more space.

For network connectivity, I did use a USB 802.11g network card but found it not to be too reliable. I now plug the frontend machine into a Linksys WRT-54GL running the OpenWRT firmware in "bridged client mode" so it connects to my wireless network. I happily run `mythfrontend` on other machines on the network. I also keep an old analog card in a machine upstairs as a remote backend. Michael also uses 802.11g wireless and has remote frontends without having any network bandwidth problems.

I used to use a WinFast TV 2000 Deluxe as a capture card—one of the cheapest analog capture cards I could find when I was first going to experiment with MythTV (and didn't want to waste my hard-earned student dollars). I now use a WinFast DTV1000 PCI card and have been working on adding support to the Linux drivers for the new revision of a Leadtek DVB USB dongle.

Michael's Hardware Choices

I chose a mini-tower Intel Pentium 4 machine, with a 3GHz processor, 80GB internal SATA hard disk for the operating system, and four USB 2 hard disks. I rapidly added another two 400GB hard disks, although this was relatively easy because of the LVM configuration I chose (which is discussed later in this chapter). My tuner is a Haupauuge PVR 350, which includes a remote control and is well supported with my analog cable TV and Linux. In hindsight, I would probably have gone for the Hauppauge PVR 500 and a USB remote control if I had thought more about it. I display the TV directly on an LCD monitor, although the Hauppauge card also supports output to a TV.

Installing the Required Software

Now you're ready to start installing the various software components of the system. In this chapter, we will cover only the dependencies for MythTV, not the installation of MythTV, which is covered in the next chapter. Chapters 4 and 5 cover the configuration of a basic MythTV system.

Linux Distribution

We are using the Ubuntu Linux distribution for all the examples in this book. It is also the distribution we both use for our MythTV boxes. Its six-month release cycle means it is up-to-date, and compiling new releases of software is often trivial. It also is based on Debian and as such gets its amazing software package management system.

We've found the installer to be easy to use and that upgrades to new versions of Ubuntu are rather painless. Also, the hardware support is excellent, and the user community in forums, mailing lists, and general writings on the Internet are excellent and have usually solved any problem you might have before you get it.

It's also worth mentioning the Fedora distribution because it also has a regular release cycle, up-to-date packages, and a good installer. However, upgrades might not be as seamless as Ubuntu, and Stewart has found its software upgrade utility leaves a lot to be desired (it's just slow, really slow).

The previously mentioned KnoppMyth, a distribution based on the popular Knoppix LiveCD, is popular among some people too because it simplifies some of the install process.

File System and Disk Layout

A few years ago it seemed fashionable to write a journaling file system (one that doesn't require a file system check after a crash) for Linux. Quickly we went from having none to having ReiserFS, ext3, JFS, and XFS. Each file system has different properties that might lead you to choose it for a specific application.

Stewart chose XFS for the following reasons:

- Designed for large disks, files, and IO loads.

- Robust and reliable.

- Near raw disk performance. There is very little file system overhead—especially for large files (your TV recordings of a few gigabytes qualify as large files).

- Fast deletes of large files (essentially instant).

- Full suite of utilities:

  - `xfscheck` and `xfsrepair`: In the event of corruption (unlikely due to journaling), the check and repair utilities work well.

  - `xfs_bmap`: This prints a list of extents that have been allocated to a file. An *extent* is how XFS tracks which space has been allocated to a file and is the number of the first block and a count. In an ideal situation, all files have only one extent. In reality, this often doesn't happen. The larger the extent, the better, because disks operate best on contiguous access.

  - `xfs_fsr`: This stands for XFS File System Repacker. This offers online defragmentation and tries to reduce the number of extents a file uses.

  - `xfsdump` and `xfsrestore`: You can back up and restore the file system efficiently. In practice, you would never use this for the file system with your media on it (who has backup media with a few hundred gigabytes of capacity at home?). However, it's useful for the file system with the operating system on it because if you do something silly (or the disk dies), getting your MythTV system back up and running can be easier.

  - `xfs_growfs`: This is the online `grow_fs`. If you want to expand the space used for recordings, you can plug in extra disk drives, and without having to unmount the file system (that is, while the computer is running and, for example, recording TV), you can grow the file system used for storing the recordings. This is most powerful with LVM and disks that can be hot-plugged (such as USB disks). Note that you can't shrink an XFS file system, and this can cause problems if you ever want to do that.

As a disclaimer, Stewart used to work for Silicon Graphics, the company that made XFS. However, he was a fan and user of XFS before he joined Silicon Graphics.

Michael, however, decided to choose the more traditional ext3 because it's the file system he uses everywhere, because the performance is fine for the workload it's being used for, because it's easy to resize, and because he's not as trusting of some of the newer file systems. ext3 is a good choice for root and other system file systems even if you're going to use XFS for the file system with your video files on it. One common criticism of ext3 is that it is slow to delete large files like those produced by MythTV. This is addressed in MythTV 0.20 with a Delete Large Files Slowly option that stops the user interface from locking up while waiting for the delete to finish.

Disk Layout, Logical Disks, and Meta Disks

The LVM allows you to group physical partitions on your disks into larger logical volumes, on which you can then create a Linux file system. It can be convenient, since the simplest MythTV configuration supports only a single directory for video recording storage; however, with power comes complexity and sometimes fragility. Figure 2-1 shows a simple LVM setup.

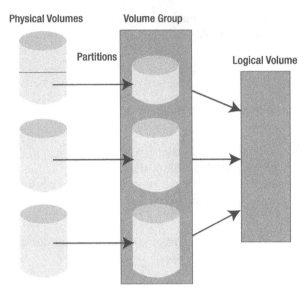

Figure 2-1. *A simple LVM setup*

Here, we have three physical disks. The first disk has two partitions, only one of which is being used as an LVM physical volume. The other partition might be for the /boot file system, the root (/) file system, or a different operating system. The other two disks are being fully used as LVM physical volumes. A *volume* group is a group of physical volumes. You allocate space from a volume group to form logical volumes. A logical volume is where you store data. Here, we're using all the space in the physical volume for one volume group. We could format this volume as an XFS file system and mount it as /home. In the future, we could add physical volumes to the volume group, extend the logical volume (using the lvextend program), and increase the size of the file system (using xfs_growfs).

You need to be careful, too, about what you call your LVM objects—they will end up as files and directories in /dev, and this can be confusing. Common naming schemes either prepend or append some combination of "vg" and "lv" to the names.

You can also build LVM volumes on top of RAID volumes. So, you have the flexibility of LVM along with the redundancy of RAID. If you don't use RAID and just have LVM, you have what's referred to as JBOD—Just a Bunch of Disks. Typically, if one of the disks in a JBOD dies, any file system that was even partly stored on that disk is gone. However, some file system repair programs (for example, xfsrepair) can "repair" a file system where one of the disks in a JBOD failed and has been replaced with another one. Typically you will lose any file whose inode (the data structure that represents a file) was stored on the lost disk, the parts of a file that were stored on the lost disk, and any directory information (filenames, although the files themselves might still exist—and be moved to the lost+found directory) that was stored on the lost disk. Running a file system repair on such a system is a "neat trick" and shouldn't be relied on for any data you care about.

MD means *multiple disk* or *meta disk* depending on who you talk to and is the software RAID implementation for Linux. There are many different levels of RAID, some even living up to the *redundant* part of what RAID stands for (Redundant Array of Inexpensive Disks):

- RAID 0 gives you no redundancy; it simply interleaves storage across two or more disks. For example, the first 4KB is stored on disk A, the second is stored on disk B, the third is stored on disk A, and so on. The purpose of RAID 0 is an increase in both read and write throughput. If either disk fails, your data is gone. This is called *striping*.

- RAID 1—or *mirroring*—gives you full redundancy at the expense of requiring twice as much disk space. Disk A and B both have the same data. Read throughput can be higher, but write speed suffers (because of the overhead of having to send the data to two disks). If one disk fails, you have full access to your data. If both disks fail, your data is gone.

- RAID 5 gives you redundancy with three (or more) drives. One drive out of the three can fail, and you still have access to your data. If two (or more) drives fail, your data is gone. So you get about two-thirds space efficiency, but it does give you redundancy.

For more information about RAID, refer to the Wikipedia description at `http://en.wikipedia.org/wiki/RAID`. The decision about how critical your TV recordings are is of course up to you. Stewart's answer is "not very." If the disk holding his TV recordings dies, he'll be annoyed, but he's not going to cry over it. Michael, on the other hand, uses mirrored disks for all his recordings, because he often has a large backlog of video to watch because of work commitments.

However, other data might be more important than your videos—for example, your photo and music collection, first because it's hard to redo all your holidays to take pictures again and second just because of the amount of time it can take to rip and encode large numbers of CDs. You might choose to store some of this data on a redundant system. Another alternative is to just store it on two PCs (on the MythTV box and on your desktop) and to make backups.

Largely, the amount of redundancy and backup you have is going to be a function of your paranoia, faith in technology, and size of your budget.

See Chapter 10 for some more information on using LVM.

Graphical Environments

Some people prefer to trim down their Linux installation to have only the bare minimum needed for MythTV. We prefer to keep a standard desktop environment. This allows you to also quickly jump on your MythTV machine to do anything you might do regularly on a computer, lets you easily use the graphical software update utility when you're in the process of setting up the machine, and lets you easily switch to a web browser, terminal, or game of solitaire.

Installing the Software

Ubuntu 6.06 is possibly the easiest operating system to install—ever. The only easier way is to have it preinstalled when you buy the computer. However, you might run into problems, which will probably be answered on the Ubuntu wiki and user forums. Always check the Internet; it's amazing what problems people have already solved.

We can't unfortunately provide complete support for your installation of Ubuntu in this book. This is because of the massive variation in hardware available today and the rapid rate at which that hardware changes.

The following sections will walk you through an Ubuntu install, step by step, and provide as much advice as possible. Remember that you can use other versions of Linux, if you're more comfortable with them. You might still find it useful to read through these sections, though, even though the exact steps will differ.

For more information about Ubuntu, you might want to check out *Beginning Ubuntu Linux: From Novice to Professional* (Apress, 2006).

Once you have assembled your hardware, you're ready to install the operating system.

Getting Ubuntu

You can download the installation CD and burn it yourself. Many Internet service providers will have an Ubuntu mirror on one of their servers, possibly not charging for bandwidth from it or providing quicker access. With the 6.06 release of Ubuntu, you can choose from three CDs. You need only one of them:

- Desktop contains a simple, graphical-based installer and components for a desktop system.

- Server contains a text-based installer, designed for installing server systems. You cannot use this version to install MythTV, because even backend-only machines require the X Window System to be installed.

- Alternate contains the more advanced installer (allowing you to configure LVM and RAID during install), and the install is text based rather than graphical. The installation procedure is, however, similar to the graphical installer.

You should also download the right CD for your architecture. This will always be x86 (unless you're constructing a rather unusual MythTV box)—you just have to work out whether your system is 32 bit or 64 bit. You can run 32-bit software including the operating system on 64-bit hardware, though. In the past some people have had problems with using the proprietary NVIDIA and ATI drivers on 64-bit operating systems but not when they run a 32-bit operating system. Keep this in mind if you are buying 64-bit hardware.

You can also buy Ubuntu DVDs from Amazon and probably from many local retailers, or you can get them for free from Canonical, the makers of Ubuntu, by visiting `https://shipit.ubuntu.com/`.

Installing Ubuntu

To begin the installation, make sure your system is set up to boot from a CD, insert the Ubuntu CD, and turn on the machine.

The first thing that loads off the Ubuntu CD is a menu letting you choose to install the system, install it in a safe graphics mode, check the CD for defects, check the memory of your PC, or boot from the hard disk (Figure 2-2). You will usually want the first option.

Ubuntu will now load, and at the end you should see a desktop. If you don't, try returning to step 1 and choosing the safe graphics mode. Ubuntu will now load the kernel, start some services, and log into a desktop environment. The boot process will take a little while to complete, but when the machine is ready for use, you should end up with a desktop that looks a lot like that shown in Figure 2-3.

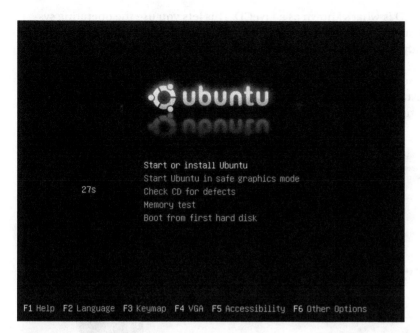

Figure 2-2. *The Ubuntu boot menu*

Figure 2-3. *The Ubuntu Install CD desktop*

The Ubuntu Install CD is what's known as a LiveCD. This lets you use the operating system and programs without installing it. You can have a look around if you haven't seen Ubuntu before. When you're ready to install, launch the installer by double-clicking the Installer icon on the desktop.

The first question you'll be asked is regarding the language you'd like to use for the installation process (and the installed system—see Figure 2-4). We'll use English, but you're welcome to choose your native language. When you've made your selection, click Forward to go to the next step.

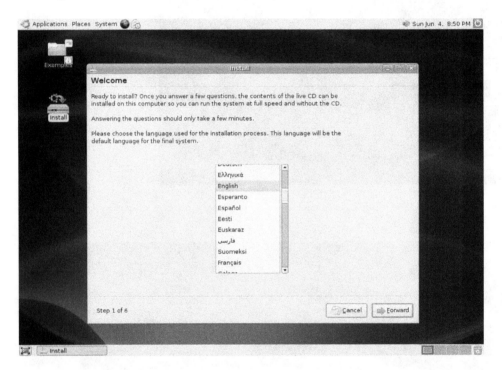

Figure 2-4. *Language selection dialog box*

You will now be asked about the current time and your time zone (Figure 2-5). When you've made your selection, click Forward to go to the next step.

The next question (Figure 2-6) is asking you to select your keyboard layout. If you're unsure, go for the one with the name of your country, or if it's a standard QWERTY keyboard, go with American English. When you've made your selection, click Forward to go to the next step.

Figure 2-5. *Selecting the time zone*

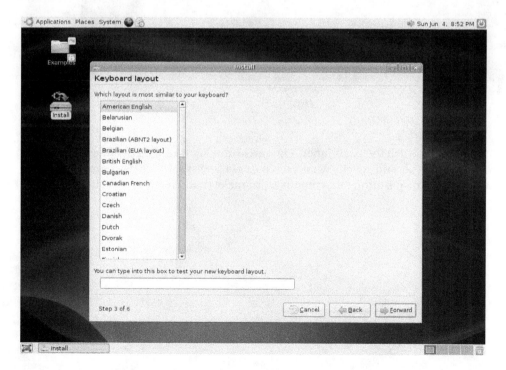

Figure 2-6. *Selecting keyboard layout*

We will now set up a user account for the MythTV box (Figure 2-7). A good way to set up the machine is with a user who automatically logs in, runs all the MythTV processes, and can run administration utilities. Although this is not the most secure way to run a Linux machine (you certainly wouldn't do it this way on a server, for example), it's extremely convenient for a machine that everybody in the family is going to use. You're not ready to set up the automatic login yet, but here is where you create the account you will use to do that. When you've made your selection, click Forward to go to the next step.

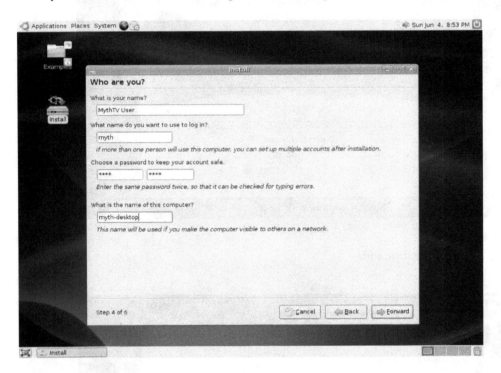

Figure 2-7. *Setting up a user account*

Ubuntu will now launch the partitioner (Figure 2-8 and Figure 2-9). This allows you to set up the layout of your disks and specify what each part of the disks will be used for. You will probably want to edit the partition table manually, so make that selection, and click Forward.

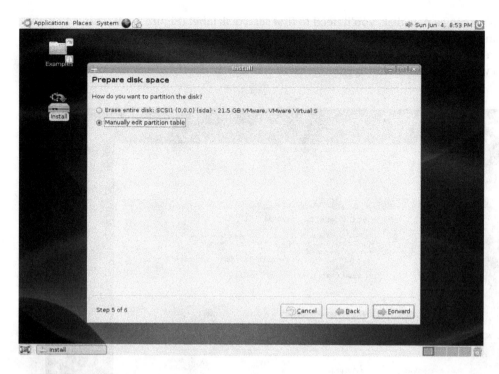

Figure 2-8. *Partitioner options*

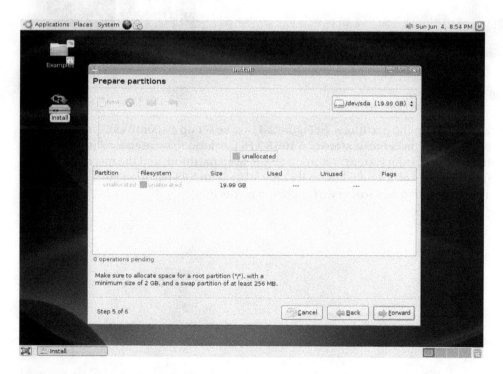

Figure 2-9. *The main window of the partitioner tool*

If you're using a new disk, you'll need to first set a disk label (Figure 2-10). This is how the partitioning information is written to disk. Standards are great; everybody has their own. So, you have a choice. Unless you feel strongly another way, you'll want the default MS-DOS disk label. You will be asked to confirm your decision.

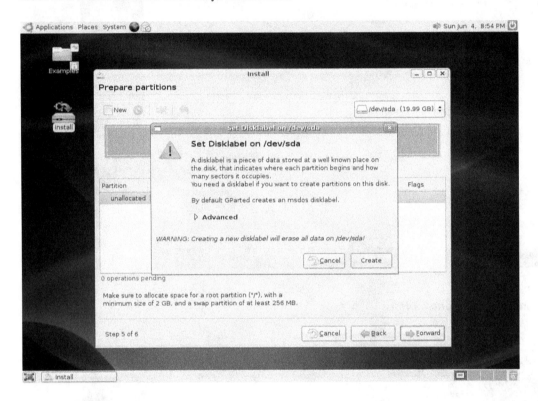

Figure 2-10. *Setting a disk*

You can now set up the partitions. In Figure 2-11 we've set up a 100MB ext3 partition to be used for /boot (where the kernel is stored); a 10GB XFS partition to be used for / (the root file system, where the operating system is stored); a 1GB swap partition; and the rest of the disk (in the example we've used a small disk, so it's only 9GB) as an XFS partition for /home (where all the user files are stored and where we'll store the media files).

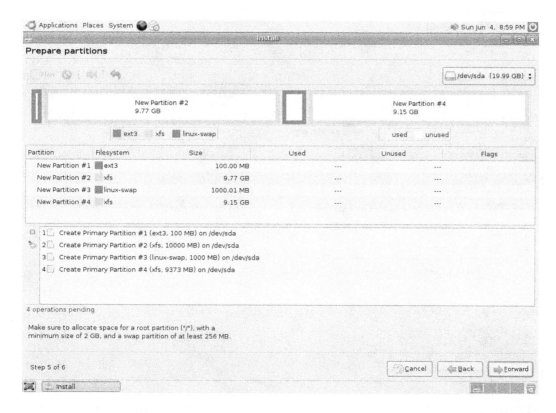

Figure 2-11. *Using the partition editor*

When you're done, click Forward. You will be asked whether you want to apply the pending operations. If you're sure you want to do this (which will destroy whatever data was on the disk previously), click Apply.

Depending how big the disk is and what file systems you chose, it could take a few seconds or several minutes to make the partitions and file systems (Figure 2-12). If you would like a drink, feel free to go get one.

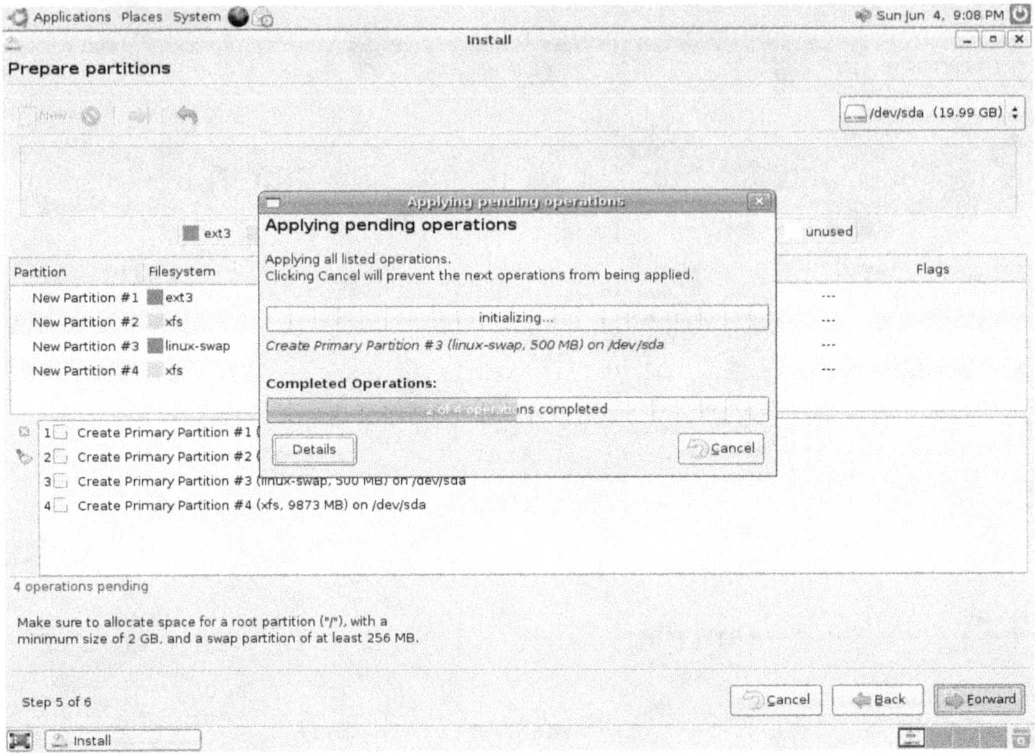

Figure 2-12. *The installer creating the partitions*

When done, the installer will ask you where you want to mount the partitions (Figure 2-13). Here we've set it up as mentioned previously.

The installer will now show you the settings you've entered before continuing with the install (Figure 2-14). When you're ready to proceed, click Forward.

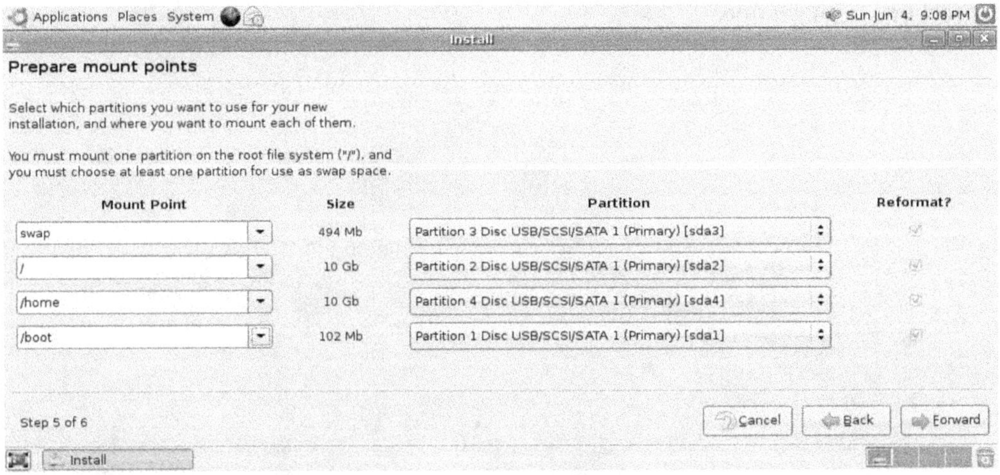

Figure 2-13. *Setting mount points for the new partitions*

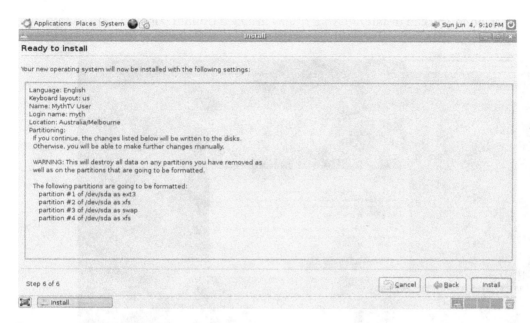

Figure 2-14. *Confirming installation options*

Ubuntu will now install (Figure 2-15). This will take a while. If you're hungry—now is the time to get something to eat. If you're not hungry, you might be by the time this is finished.

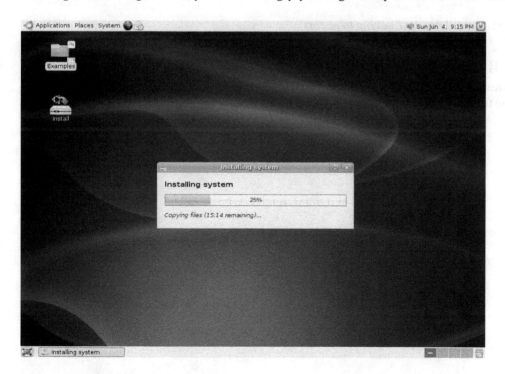

Figure 2-15. *Installing the system*

When done, you can restart into your freshly installed Ubuntu (Figure 2-16).

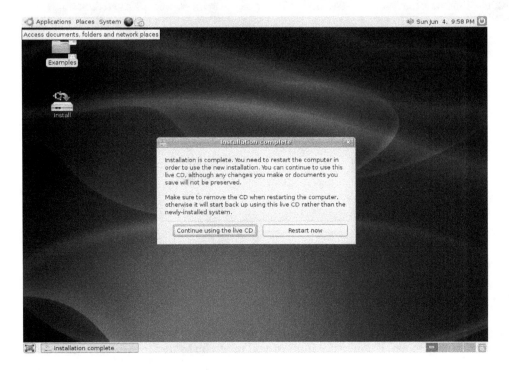

Figure 2-16. *Install finished, asking to restart*

Setting Up Automatic Login

When your system has finished booting, you'll be presented with the login screen (Figure 2-17). Enter the username and password you set up in the installer to log in.

Figure 2-17. *Ubuntu login screen*

After you've logged in, you can set up the autologin. Open the Login Window configuration program (Figure 2-18).

Figure 2-18. *Launching the Login Window configuration program*

You will need to enter your password to perform this administrative task (Figure 2-19). This is the same password as your login password (the one you set up in the installer). Many of the administrative tools require you to enter your password. This is a graphical interface to the sudo command (superuser DO)—which gives you superuser privileges for running a command. Users who can run sudo (and what commands they can use with it) are listed in the /etc/sudoers file.

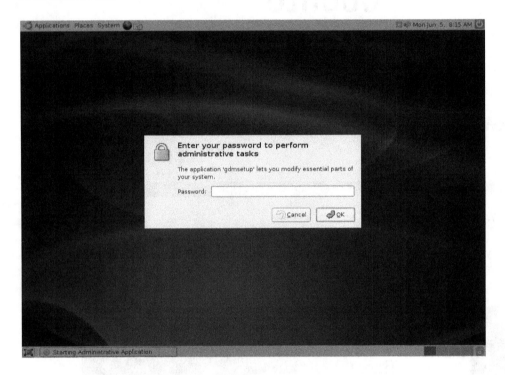

Figure 2-19. *Many administrative tasks require you to enter your password.*

Now the Login Window tool will launch (Figure 2-20). On the Security tab, enable Automatic Login, and select your user. When done, click Close.

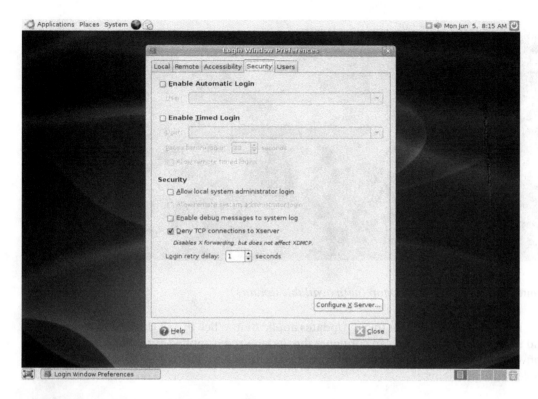

Figure 2-20. *The Login Window administrative tool*

Installing Software Updates

When software updates are available, an orange star will appear in the notification area along with a little pop-up message (Figure 2-21). It's quite normal for there to be updates waiting when you first install a Ubuntu system. Click the orange icon to launch the Software Updates application.

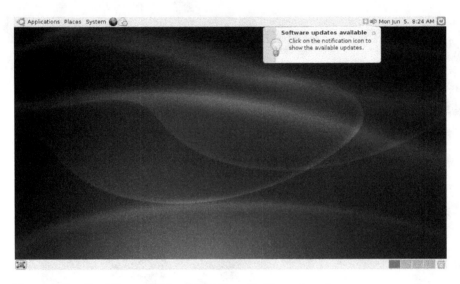

Figure 2-21. *The Ubuntu desktop (with available updates)*

Figure 2-22 shows the Software Updates application. Click Install Updates to download and install the updates. You can configure the updates to be applied automatically from the Software Properties Administration utility, although there is a small possibility that an upgrade could cause problems.

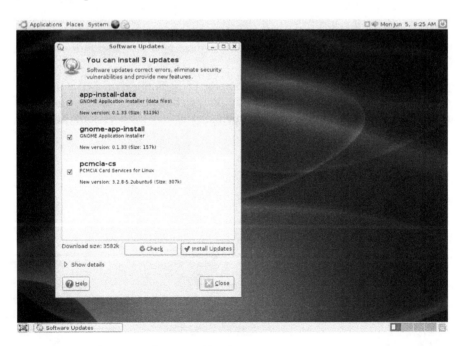

Figure 2-22. *Software Updates application*

Verifying That Your Hardware Works

Now that Ubuntu is installed and you have the updates applied, it's a good idea before going much further to test that your hardware works as you expect it to work.

General Issues to Consider

If you have a problem with any of the following, you can look in a few places to help track down the problem.

The first place is the output of the dmesg command after start-up. This shows the most recent log messages from the Linux kernel. During start-up or when any driver is loaded, messages are often printed here saying "Detected X" (where *X* is a hardware device). You can scan through these messages for anything related to your hardware. For example, successful detection of a Nova-T PCI Digital TV card can look like this:

```
DVB: registering new adapter (TT-Budget/WinTV-NOVA-T  PCI).
tda1004x: Detected Philips TDA10045H.
tda1004x: Detected Philips TDM1316L tuner.
DVB: registering frontend 0:0 (Philips TDA10045H)...
```

For a wireless card in one of our laptops, a message like this appears:

```
ipw2200: Detected Intel PRO/Wireless 2200BG Network Connection
ipw2200: Detected geography ZZM (11 802.11bg channels, 0 802.11a channels)
```

The format of these messages will likely be different for each driver and perhaps between versions of the kernel. If you get an error message, perhaps you need a newer driver or a patch. We strongly recommend you buy hardware that is easily supported by the kernel you are running (most likely the kernel that ships with Ubuntu, which with 6.06 LTS is based on 2.6.15). Any money you might save by going for possibly cheaper but less well-supported hardware can bring you frustration when it comes to setting it up.

The other utility to look at the output of is lspci. This "LiSts PCI" devices in the system. If a device isn't listed here, then it hasn't been found on the PCI bus (the cards you plug into slots inside the computer). Some output might look like this:

```
0000:00:1e.2 Multimedia audio controller: Intel Corporation ➡
82801FB/FBM/FR/FW/FRW (ICH6 Family) AC'97 Audio Controller (rev 04)
0000:01:00.0 VGA compatible controller: ATI Technologies Inc M24 1P ➡
[Radeon Mobility X600]
```

which are a sound card and a graphics card. For USB devices, there is lsusb for "LiSt USB." You can get output like this:

```
Bus 005 Device 001: ID 0000:0000
Bus 001 Device 003: ID 077d:0223 Griffin Technology IMic Audio In/Out
Bus 001 Device 002: ID 046d:c00f Logitech, Inc.
Bus 001 Device 001: ID 0000:0000
Bus 004 Device 001: ID 0000:0000
Bus 003 Device 001: ID 0000:0000
Bus 002 Device 001: ID 0000:0000
```

The `Logitech, Inc.` device in this output is actually a mouse. Sometimes the names listed aren't the most obvious. You can always unplug a device and see what changes, though. For example, if we unplug the mouse and run `lsusb` again, the output is as follows:

```
Bus 005 Device 001: ID 0000:0000
Bus 001 Device 003: ID 077d:0223 Griffin Technology IMic Audio In/Out
Bus 001 Device 001: ID 0000:0000
Bus 004 Device 001: ID 0000:0000
Bus 003 Device 001: ID 0000:0000
Bus 002 Device 001: ID 0000:0000
```

With PCI cards you will have to shut down the machine first before you remove them, but you can happily plug and unplug USB devices, and everything should "just work."

If you didn't get any mention in `dmesg` but you did in `lspci` or `lsusb`, it might be that the driver for the device either isn't loaded or isn't installed on your system.

Sound

Ubuntu plays a start-up sound as you log in. If you heard it, odds are your sound output works. Believe it or not, we've found the most common problems for not hearing anything are the volume isn't turned up, the speakers aren't turned on, or we've plugged the speaker cable into the wrong socket. This is especially true because some sound drivers on Linux default to starting up with the volume turned all the way down. You can also play various sounds under System ➤ Preferences ➤ Sound. You can also turn off the login sound from this Control Panel, which can be good to do if you don't want to hear it every time you boot the computer.

Video Capture

You can do a simple test with some basic TV-viewing programs. The best way to do this depends on what type of capture card you have, so we'll discuss some of the more common options in the following sections.

Installing IVTV Capture Cards

First, download the latest version of the IVTV driver that supports your kernel version from `http://www.ivtvdriver.org`. At the time of this writing, the latest version is 0.7.0, which has a list of supported cards at `http://ivtvdriver.org/index.php/Supported_hardware`. If you need to determine the version of the Linux kernel running on your machine, run this command:

```
$ uname -r
2.6.15-26-386
```

Here you can see that we're running Linux 2.6.15-26 on this machine. If you're running a different version, then you will need to change the version numbers in the following instructions to match the version you are running:

```
$ sudo apt-get install gcc-3.4 linux-source-2.6.15 linux-headers-2.6.15-26-386
...
$ cd /usr/src/
$ tar --bzip -xf linux-source-2.6.15.tar.bz2
```

```
$ ln -s linux-source-2.6.15 linux
$ ln -s /usr/src/linux /lib/modules/2.6.15-26-386/build
```

If you get an error about /lib/modules/2.6.15-26-386/build already existing, then just remove it, and run the command again. Now edit the EXTRAVERSION line in /usr/src/linux/ Makefile to match the last two fields of the Ubuntu kernel version. Our Makefile ends up looking like this:

```
VERSION = 2
PATCHLEVEL = 6
SUBLEVEL = 15
EXTRAVERSION = -26-386
NAME=Sliding Snow Leopard

ifdef UBUNTUBUILD
EXTRAVERSION =
endif


# *DOCUMENTATION*
[...snip...]
```

Now you need to configure the kernel source so the build will work properly:

```
$ cp /boot/config-2.6.15-26-386 /usr/src/linux/.config
$ make oldconfig
```

Accept the defaults for anything you are prompted for here, because this means that functionality wasn't available when the current kernel was built. You select the default by just hitting Enter:

```
$ make prepare0
$ make scripts
```

Next, change directories to the place you downloaded the IVTV driver to, and execute these commands (which assume you're using the 0.7 version of the IVTV driver):

```
$ tar xzf ivtv-0.7.0.tar.gz
$ cd ivtv-0.7.0/
$ make
$ make install
$ depmod -a
```

Now the kernel driver is installed and ready to go. You do need to install the firmware that Ubuntu will load for the driver when the driver starts. This firmware is in two ZIP files that you download and then copy to the right place:

```
$ cd utils
$ wget ftp://ftp.shspvr.com/download/wintv-pvr_150-500/inf/pvr_2.0.24.23035.zip
$ wget ftp://ftp.shspvr.com/download/wintv-pvr_250-350/inf/pvr_1.18.21.22254_inf.zip
$ unzip pvr_2.0.24.23035.zip
$ ./ivtvfwextract.pl pvr_1.18.21.22254_inf.zip
$ cp HcwMakoA.ROM /lib/firmware/v4l-cx25840.fw
```

```
$ cp HcwFalcn.rom /lib/firmware/v4l-cx2341x-enc.fw
$ mv /lib/modules/ivtv-fw-dec.bin /lib/firmware/
$ mv /lib/modules/ivtv-fw-enc.bin /lib/firmware/
$ cp ../v4l-cx2341x-init-mpeg.bin /lib/firmware
$ ln -s /lib/firmware/ivtv-fw-dec.bin /lib/firmware/v4l-cx2341x-dec.fw
```

You're nearly done. All you need to do now is add these two lines to /etc/modprobe.d/ aliases:

```
alias char-major-81-0 ivtv
alias char-major-81-0 ivtv
```

You can test the results by running these commands:

```
$ depmod
$ modprobe ivtv
$ dmesg
```

Look for error messages in the dmesg output, which might indicate that something went wrong. If you need more information about this install process, check these URLs: http:// www.mythtv.org/wiki/index.php/Ubuntu_Dapper_Installation#Install_IVTV_drivers and http://ivtvdriver.org/index.php/Howto:Ubuntu. The previous description is merely an updated version of the content of those two URLs.

Testing IVTV Capture Cards

IVTV-based cards output MPEG-2 compressed video, which makes testing the card quite easy. Just use ivtv-tune to select a channel, and then you can just dump the output of the video card straight into an MPEG file. Here's an example in which we get the usage message for ivtv-tune, then tune to a channel (in this case Discovery Channel on Comcast in Mountain View, California), and finally create an MPEG file:

```
$ ivtv-tune
ivtv-tune 1.50

Channel/Frequency changer for V4L2 compatible video encoders

Usage: ivtv-tune [OPTIONS]...

  -h, --help              Print help and exit
  -V, --version           Print version and exit
  -c, --channel=CHANNEL   set new channel
  -d, --device=DEVICE     set video device node
  -f, --frequency=FREQ    set new frequency (MHz)
  -l, --list-channels     list all channels and their frequencies
                            (default=off)
  -L, --list-freqtable    list all available frequency mappings  (default=off)
  -t, --freqtable=STRING  set frequency map to use
  -x, --xawtv=CHANNEL     set new channel using custom map from ~/.xawtv
$ ivtv-tune -c 29
$ cat /dev/video0 > /tmp/video.mpg
```

Note that the name of your video capture device file will be different if you're testing the second tuner on your card (if it has one) or if you have other video devices already loaded. You can see what video device to use from the dmesg output earlier. Now, to see how your video turned out, just play it with your favorite video player. Here is an example using xine:

```
$ xine /tmp/video.mpg
```

Installing Other Analog Capture Cards

Some analog cards ship with up-to-date drivers already installed by Ubuntu. This makes testing them pretty trivial. If you're using a BT878 or any other card that uses the bttv driver, then this is the case. You can skip straight over the installation details and head to the "Testing Digital Capture Cards" section. If you need to install a driver, then you should find the download and installation links for that driver in the MythTV video capture card guide at http://www.mythtv.org/wiki/index.php/Video_capture_card. If you're installing an IVTV-based card, like many of the Hauppauge cards, then the installation instructions are relatively simple.

Testing Other Analog Capture Cards

The simplest, oldest, and least pleasant to use TV capture application is xawtv. You can install it using the Synaptic Package Manager (System ➤ Administration ➤ Synaptic Package Manager). It should automatically detect your card.

You will want to set up an .xawtv file, the configuration file for xawtv. You can also make these settings from the GUI. Our .xawtv for free-to-air Australian TV is as follows. You can see complete help for the .xawtv file by looking at the xawtvrc man page.

```
[global]
freqtab = australia

[ABC]
channel=2
fine=8

[Seven]
channel=7

[Nine]
channel=9

[Ten]
channel=10

[SBS]
channel=28

[C31]
channel=31
```

With this configuration file, we can run xawtv and check that we're getting TV reception.

Installing Digital Capture Cards

If your digital card is supported by Ubuntu, the installation procedure is rather painless; you should just see log lines in the output of the dmesg command. If your card is supported only in a newer version of the Linux DVB drivers, you might need to build the latest version of them. For the latest instructions on doing that, head to http://www.linuxtv.org.

Sometimes cards might require a firmware file that the driver downloads to the card, much like the IVTV driver we described earlier in this chapter. If something extra is required, the log message in the dmesg output will likely tell you exactly which website you need to visit to download the firmware and where to put it on your machine.

Testing Digital Capture Cards

You can use xine to view digital TV broadcasts. In the controls, you'll see a button labeled DVB (which stands for Digital Video Broadcasting). You'll want to click this. However, you need to use an external utility to change the channel. These are scan and tzap. If you are using a DVB-S card, it's szap. For DVB-C, it's czap. You will need to install the dvb-utils package to get these programs.

You first want a frequency file. You'll find quite a few in /usr/share/doc/dvb-utils/ examples/scan/; you can probably find one for your area. If not, it's easy to create one with the frequencies most likely listed on the website of the regulator in your area. You can now run scan to generate a channels.conf file:

```
$ scan /usr/share/doc/dvb-utils/examples/scan/dvb-t/au-canberra > channels.conf
```

After you've run scan, you can now create a directory under your home directory called .tzap, .szap, or .czap (depending on what's appropriate for your card). You should copy the channels.conf file there. To change the channel, you use the zap utility (that is, tzap):

```
$ tzap "ABC"
```

You should now be able to use xine and the DVB button to view the MPEG-2 stream that is being broadcast.

If you cannot tune in a TV station, your aerial might not be connected properly (or might not be adequate for digital). If you're really unlucky, like Stewart was, the TV card you have is a newer (incompatible) revision of a model that has previously worked OK, and the driver didn't give you any error. If you're slightly luckier, there might be an error in the log. Check the output of the dmesg command for some hints.

TV-Out from Your Video Card

If you have a DVI or VGA input plug on your TV, you're in luck; you've just been saved a bunch of pain. Otherwise, you might have to go through a few steps before TV-out through a S-Video or composite connector will work.

TV-Out Using the NVIDIA Driver

You will typically need to add two options to the /etc/X11/xorg.conf file and reboot to enable TV-out from an NVIDIA card using the binary-only drivers. Add the following three lines to the "Device" section that corresponds with your video card:

```
Option "ConnectedMonitor" "TV"
Option "TVStandard" "NTSC-M"
Option "TVOutFormat" "SVIDEO"
```

You should change the TVStandard setting to one matching the standard used in your country (NTSC-M is correct for the United States; a complete list is at http://www.ee.surrey.ac.uk/ Contrib/WorldTV/broadcast.html) and change the TVOutFormat setting to Composite or SVIDEO, depending on what you are using. The documentation with the NVIDIA driver has a complete list of settings.

It is useful to use the nvidia-settings program to change some options for TV-out from the NVIDIA card. A common one to tweak is overscan. When you can see the desktop on your TV set, you're done. With LCD and plasma TV sets, you might need to explicitly set a resolution for the output to get a clearer picture.

TV-Out Using the ATI Driver

Not all ATI graphics cards are able to use the TV-out functionality with the free drivers. ATI provides proprietary drivers for Linux that support most functionality of the newer cards—but not everything. It is a good idea to check online beforehand. The Ubuntu package xorg-driver-fglrx provides the driver. You need to configure it first, though.

Two programs will help you configure your ATI card: aticonfig and fireglcontrol (in the fglrx-control package). The aticonfig program is a command-line program that helps write the xorg.conf file (found in /etc/X11/), and fireglcontrol is a graphical control panel for configuring the proprietary ATI driver.

First use aticonfig -initial to generate an xorg.conf file, then copy it to /etc/X11/ xorg.conf (which will require root privileges, so use sudo cp xorg.conf /etc/X11/), and finally reboot. After that, you can use aticonfig and fireglcontrol to modify your configuration for your TV output.

Conclusion

Congratulations, you've made some hardware choices, installed the operating system, and done some basic configuration and testing of your hardware. One of the trickiest parts is now over, and you probably didn't have much trouble. You are now ready to install MythTV. Read on for details of how to do that and then for information about configuring MythTV and getting the remote control working.

CHAPTER 3

■ ■ ■

Installing MythTV

In this chapter, we will show how to install the dependencies for MythTV such as the MySQL database; `lirc`, which is the remote control driver; and then MythTV itself. We will show how to perform some basic configuration, and by the end of the chapter you should be able to watch Live TV, pause it, and change the channel. By now, you should have installed Ubuntu Linux, as detailed in the previous chapter, and done some basic checks that your hardware works.

We'll show how to build MythTV from source instead of using existing packages for several reasons: to get the right optimizations for your CPU and to be sure you get the most current stable version. If you want to run the absolutely latest code, then you should check out Chapter 15, which describes how to install development branches of the code.

Downloading MythTV

First download the source code for the latest released version of MythTV. You won't compile it just yet, but you need the code for some of the other installation steps, so it is good to download it early. The latest version of MythTV is currently 0.20, although since the time of this writing, a newer version might have been released.

Why aren't we using Ubuntu packages? In our experience, downloading from source will give you a more recent version of MythTV, and that gives you many improvements to the application (each release is consistently better). Additionally, you'll get code that is optimized for your local system. Nothing is inherently wrong with the Ubuntu packages, however, so if you would rather install from packages, then feel free to do so.

It is our experience that building and installing MythTV does not vary too greatly between releases, so installing a newer version might be just as easy for you. However, this book is based on the 0.20 release, so if in doubt, you can always stick to 0.20 and upgrade later. Generally, we recommend you run the released version because it typically will be more stable.

To download the source, go to `http://www.mythtv.org`. You'll see a Downloads link on the left side of the page. Click that, and you'll see a page that contains a MythTV link in the middle of the page. Click that link. You'll then see a list of three downloads: the MythTV code, the code to MythTV's plug-ins, and a set of MythTV themes. Download all three.

If you are feeling adventurous, you can instead run the development version of MythTV, which is covered in Chapter 15.

Next you need to decompress and extract the files you downloaded into their source code. Open a terminal window, and then change directories to where you saved the MythTV sources.

For example:

```
$ cd mythtv-src
```

Now, extract the MythTV source code from the `tar.bz2` file. In our case, we downloaded the source code for 0.20, which means we have the following filenames. They might be different if you downloaded a different version of MythTV. Extract the source code like this:

```
$ tar xfj mythtv-0.20.tar.bz2
$ tar xfj mythplugins-0.20a.tar.bz2
$ tar xfj myththemes-0.20.tar.bz2
```

The directory should now look something like this (the `ls` command lists what's in the current directory):

```
$ ls
mythplugins-0.20a          myththemes-0.20          mythtv-0.20
mythplugins-0.20a.tar.bz2  myththemes-0.20.tar.bz2  mythtv-0.20.tar.bz2
```

Installing Prerequisites

Before you can build and install MythTV, you need to install some required software and libraries. As with the other instructions you'll find in this book, we will show you how to install these prerequisites utilizing the tools within Ubuntu. If you've chosen to utilize a different operating system, the methods to complete the installs might differ.

Using apt-get

Ubuntu (like the Debian Linux distribution it is derived from) has a program called `apt-get` whose role is to help you install software. It does this by managing what dependencies (other pieces of software) are needed to run the piece of software you are installing and by installing those at the same time. It also keeps a database of all the software it has installed, which makes upgrading much easier later. It's a command-line program, and we'll show how to use it to install needed software throughout this book. You can also use the graphical interface to `apt-get`; it's called Synaptic Package Manager, and you can find it in System ➤ Administration.

For more information about Ubuntu packaging, check out one of the many excellent Ubuntu or Debian books that are available, or refer to some of the tutorial sites available for those distributions.

Setting Up the MySQL Database Server

MythTV stores its recorded programs, other digital video and audio files, and themes on the Linux file system; all other information, it keeps in a SQL database. This includes guide data downloaded from the Internet, the programs you have scheduled to record, the name of the theme you've selected for each graphical frontend, the programs you previously have recorded, and the TV channels you can receive as well as what they're called. MythTV uses the MySQL database server to store this information. Luckily, MySQL is really easy to set up and get running with MythTV.

To install MySQL, you run this:

```
$ sudo apt-get install mysql-server
```

After the package is installed, you'll have a running MySQL server with a set of default options that will be fine for your MythTV setup.

Setting Up the Database Schema

You can now insert the default data into the MySQL database that MythTV will use. This includes setting up the schema, which is the layout of the tables that the data is stored within inside the database. You do this by running a little SQL script, which is provided with the MythTV source code, in the database subdirectory:

```
$ mysql -u root < mythtv-0.20/database/mc.sql
```

This sets up the database so that only connections from the machine on which MySQL is running will work. If you are going to run remote frontends, you'll need to change this (you can use the graphical MySQL Administrator to make this easy). MySQL Administrator is installable using apt-get or Synaptic in the mysql-admin package; you can find it in Applications ➤ System Tools.

If you intend to make your MySQL server available to anyone other than the local machine, you should seriously consider setting a password for the root MySQL account as well. This is because this account has administrative access to the database server and can do things such as delete entire databases—you obviously don't want random people doing that. You can find more information about setting a root password at http://dev.mysql.com/doc/refman/5.0/en/passwords.html.

Getting the Libraries Needed to Compile MythTV

You'll need some extra software packages installed to compile MythTV. These include KDE (http://www.kde.org) development libraries along with libraries used for accessing TV capture cards. We're again going to use apt-get to install software for us—in this case, the build dependencies (software needed to build this software) for MythTV. The command to use is as follows:

```
$ sudo apt-get build-dep mythtv
```

This will search for all the software needed and ask whether you want to install it (you do). Some software might need to be downloaded from the Internet, so be patient. However, this command is going to install a lot of stuff, so it seems useful to show some sample output. This is what we get when we run the command on a machine that has only just had Ubuntu installed on it:

```
$ sudo apt-get build-dep mythtv
Reading package lists... Done
Building dependency tree... Done
The following NEW packages will be installed:
  binutils build-essential cpp cpp-3.4 cpp-4.0 debconf-utils debhelper dpkg-dev ➥
g++ g++-3.4 g++-4.0 gcc gcc-3.4 gcc-3.4-base gcc-4.0 html2text libartsc0 ➥
libartsc0-dev libasound2-dev libaudio-dev libc6-dev libdvb-dev libexpat1-dev ➥
libfontconfig1-dev libfreetype6-dev libgl1-mesa-dev libglib2.0-dev ➥
libglu1-mesa-dev libice-dev libjpeg62-dev liblame-dev liblame0 liblcms1-dev ➥
liblircclient-dev libmng-dev libmysqlclient14 libmysqlclient14-dev libogg-dev ➥
libpng12-dev libqt3-headers libqt3-mt libqt3-mt-dev libsm-dev libstdc++6-4.0-dev ➥
```

```
libstdc++6-dev libvorbis-dev libx11-dev libxau-dev libxcursor-dev libxdmcp-dev ➥
libxext-dev libxfixes-dev libxft-dev libxi-dev libxinerama-dev libxmu-dev ➥
libxmu-headers libxrandr-dev libxrender-dev libxt-dev libxv-dev libxvmc-dev ➥
libxvmc1 libxxf86vm-dev linux-kernel-headers make mesa-common-dev ➥
qt3-dev-tools x-dev x11proto-core-dev x11proto-fixes-dev x11proto-input-dev ➥
x11proto-kb-dev x11proto-randr-dev x11proto-render-dev x11proto-video-dev ➥
x11proto-xext-dev x11proto-xf86vidmode-dev x11proto-xinerama-dev zlib1g-dev
The following packages will be upgraded:
  libpng12-0
1 upgraded, 80 newly installed, 0 to remove and 63 not upgraded.
Need to get 38.0MB of archives.
After unpacking 135MB of additional disk space will be used.
...
```

You can see that this command installs quite a few packages if they're not already installed. You will also need the MySQL module for QT (a library that MythTV uses). You need it so that MythTV will be able to talk to the MySQL database. Install it with the following:

```
$ sudo apt-get install libqt3-mt-mysql
```

Setting Up the Shared Libraries

We're going to install MythTV in the /usr/local directory, which keeps things separate from the other packages installed on the system. People familiar with Linux systems might choose to install MythTV elsewhere, or if you've decided to use some prebuilt packages, the decision will have been made for you. You need to tell the system that it might need to look in /usr/local for various libraries that you'll install. To do this, run the following commands:

```
$ sudo -s
# echo /usr/local/lib >> /etc/ld.so.conf
# /sbin/ldconfig
# exit
```

Setting Up Your Remote Control

If TV was meant to be interrupted by getting out of your chair for a cold beer, there wouldn't be armchairs with built-in fridges. Before you can compile the MythTV code, you need to decide what form of remote control you are going to use, if any. This is because you need to configure some of the remote control methods before you compile MythTV to work. Four remote control options are available with MythTV:

- You don't have to use a remote control. You can choose to use a keyboard (which might be a Bluetooth wireless keyboard) to control MythTV. This requires no setup.

- You can use an infrared keyboard and a learning remote. You can teach the remote control keyboard codes and then just pretend a keyboard is plugged in. Apart from teaching the remote control the right keycodes, this needs no setup.

- You can use lirc to implement a full-blown remote control. This requires a fair bit of setup, and you need to do it before you compile MythTV. We will discuss this option extensively.

- You can use the MythTV network control interface to control MythTV from a computer or PDA. In fact, you can also use Jabber and Google Talk instant messages to control MythTV via this interface. We discuss this in Chapter 13.

Setting Up a Remote Control with IR Keyboards and a Learning Remote

Many people decide that setting up lirc is either too hard or too much of a hassle, or they decide they also want a wireless keyboard (for things such as a web browser) and a universal remote that talks to all their gadgets in their living room. One trick is to buy an infrared keyboard and teach a learning remote certain keystrokes for various buttons on the remote control. This lets you use the keyboard or the learning remote to control MythTV.

It's important to note that finding a good infrared keyboard can be difficult these days because they are far less common now that radio frequency and Bluetooth keyboards are more affordable and popular.

Setting Up a Remote Control with lirc

The software package that will help you set up a complete remote control for MythTV is lirc. lirc is the Linux Infrared Remote Control package. It is flexible, which is really just a way of saying that it can be quite complex. Although you can use a wireless keyboard as a remote for MythTV (and many people find that more convenient), here we're talking about something that looks like a traditional TV/VCR handheld remote.

Many TV capture cards come with a remote control and infrared receiver that plugs into the TV card (see Figure 3-1 for some examples). You can also get USB remote controls quite easily that are well supported with lirc. This is a good, cheap, and easy way to get a remote control for your MythTV system.

You'll now learn how lirc works, make sure MythTV and Xine have support for lirc, install the latest version of lirc, and create the configuration files for your remote control.

Figure 3-1. *Three remote controls shipped with TV capture cards and an IR sensor*

■**Caution** lirc can be difficult to configure and get going exactly right. The reward of having an efficient-to-use remote is significant, though. Stewart spent about two hours getting everything set up exactly right. Your mileage might, of course, vary.

Understanding How lirc Works

The job of lirc is to translate the infrared signals from the remote into commands and make them available to programs running on your machine. A kernel module interfaces with the hardware and provides a /dev/lirc device. A program called lircd (the lirc daemon) runs in the background and provides an interface for applications. You can use this interface in one of two ways. There is a program called irxevent that simply turns the remote control button presses into X Window System events (exactly like the keypresses or mouse movements that keyboards and mice generate). The only problem with this is if you need to map buttons on the remote to different keys in different applications (for example mythfrontend and xine).

The mapping between buttons on the remote and what they do is defined in the lirc configuration file—.lircrc, which is stored in the home directory. This is an example of some of the settings in the .lircrc file:

```
begin
     prog = mythtv
     button = CH_UP
     repeat = 4
     config = Up
end

begin
     prog = mythtv
     button = CH_DOWN
     repeat = 4
     config = Down
end

begin
     prog = mythtv
     button = VOL_UP
     repeat = 4
     config = Right
end
```

The alternate—and ideal—way is to have each application you want to use the remote control with built directly with lirc support (in other words, have the applications compiled to include the lirc libraries needed to work with lirc). Both MythTV and Xine support this, and the xine that comes with Ubuntu is already compiled with lirc support.

Installing lirc

We'll now show you the process of how to install `lirc` so that we can use the remote control that came with our TV capture card. Using other supported "remote dongles" (just an IR receiver, not attached to a TV tuner) is a similar experience. We've noted when there are differences between the two.

We have always installed `lirc` from source code, not from the packages provided with a Linux distribution (such as Ubuntu). This is to ensure that we get a more up-to-date, compatible, and (theoretically) less buggy system. Both authors have experienced problems with getting older versions of `lirc` to work correctly with their remote controls. New versions from the website have always "just worked," though.

The first step is to download the latest version of the `lirc` source code from http://www.lirc.org/. At the time of writing, this was 0.8.1. Next, ensure you have the dependencies needed to build `lirc` installed. You will also need to install the `dialog` package, which isn't automatically part of the build dependencies, so install it too:

```
$ sudo apt-get install dialog
$ sudo apt-get build-dep lirc
```

You also need to make sure you have the kernel headers installed and prepared (replace `VERSION` and `RELEASE` with what you worked out by reading the "Finding Your Current Kernel Version" sidebar). Follow these steps to install and build the kernel headers. It is important to follow these steps and not just use the Ubuntu `linux-headers` package, because it does not have all the required files for building `lirc`. You might have to redo these steps of installing `lirc` after certain updates to the kernel (the symptom being your remote control not working after a kernel upgrade).

```
$ sudo apt-get install linux-source-VERSION
$ cd /usr/src
$ sudo tar xfj linux-source-VERSION.tar.bz2
$ sudo ln -s /usr/src/linux-source-VERSION /lib/modules/RELEASE/build
$ cd linux-source-VERSION
$ cp /boot/config-RELEASE .config
$ sudo make scripts prepare-all
```

FINDING YOUR CURRENT KERNEL VERSION

The `uname` program will tell you which version of the kernel you are running. Running `uname -r` from a terminal will give you output like "2.6.15-27-686." This is the full release "name" (the `-r` is for release) of the kernel you are running. The actual meaning of the numbers isn't significant here. However, the version is just the first three numbers (2.6.15), and this is the `linux-source-VERSION` that you'll need to install.

Extract the source code on your MythTV system:

```
$ tar xfj lirc-0.8.1.tar.bz2
```

The j is for a file ending in `.bz2` .If yours ends in `.gz`, replace the j with a z. This will create a directory named `lirc-0.8.1` that contains the source code for `lirc` version `0.8.1`.

Go into the directory that contains the source, and run the `lirc` setup script:

```
$ cd lirc-0.8.1
$ ./setup.sh
```

You'll now see a menu. Use the arrow keys to navigate, and press Enter to select. The first option is selecting the driver you want to use. Under this menu, you'll see a variety of categories. Since we're going to use the remote that came with our TV card, we'll talk about that menu. If you're using another kind of remote control, then you'll need to specify the right option here for your specific remote (see Figure 3-2).

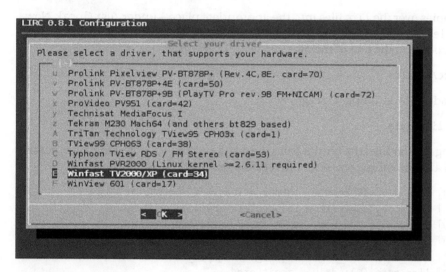

Figure 3-2. `lirc` *setup*

With luck, your TV card (or something similar to it) will appear in this menu. If it's not obvious, look at the `lirc` table of supported devices at `http://www.lirc.org/html/table.html` (keep in mind that this can be out-of-date, so you might need to resort to web searches). Now that you've selected the driver, you can click Save Configuration & Run Configure. This is developer talk for "go to the next stage of installation." The configure step should go flawlessly. If there's an error, make sure you have the prerequisites installed properly. At the end of the configure stage, it will tell you which kernel module you will be using. Write this down because you will need to add this to the `/etc/modules` file. Here is an example of the output:

```
...
config.status: creating drivers/lirc_sir/Makefile
config.status: creating drivers/lirc_streamzap/Makefile
config.status: creating daemons/Makefile
config.status: creating tools/Makefile
config.status: creating doc/Makefile
config.status: creating doc/man/Makefile
config.status: creating config.h
```

```
config.status: config.h is unchanged
config.status: executing default-1 commands

You will have to use the lirc_gpio kernel module.

Now enter 'make' and 'make install' to compile and install the package.
```

You can see here that this remote control uses the `lirc_gpio` kernel module. Next, run `make` to build `lirc`, and then install the built code:

```
$ make
$ sudo make install
```

The install copies over the `lirc` files into the appropriate place—under `/usr/local`. Now edit the `/etc/modules` file, and add the module mentioned earlier. That's as simple as just adding the name of the module to the end of the file:

```
$ sudo gedit /etc/modules
```

Install the module (for example `lirc_gpio`):

```
$ sudo modprobe lirc_gpio
```

After installing the module, the output from the `dmesg` command should indicate that the driver was loaded and something was either detected or enabled:

```
[17282870.488000] lirc_dev: IR Remote Control driver registered, at major 61
[17282972.708000] The bttv_* interface is obsolete and will go away,
[17282972.708000] please use the new, sysfs based interface instead.
[17282972.708000] lirc_gpio (-1): card type 0x22, id 0x6606107d
[17282972.708000] lirc_dev: lirc_register_plugin: sample_rate: 0
[17282972.712000] lirc_gpio (0): driver registered
```

Now you can test and configure `lirc` so that you can use your remote with MythTV.

Using Remote Control Dongles: Remotes Not Connected to a TV Card

You'll find yourself shopping for a remote control dongle if you have a frontend without a TV tuner or you want a different remote control from the one provided with your TV tuner. You can find the current list of supported dongles at http://www.lirc.org/ (at the time of writing, it is on the right, and you have to scroll down to see it). We recommend the Streamzap PC Remote Control device, which costs about $30 US on Amazon.com.

Testing Your lirc Installation

For this initial test, you'll first check that the driver is working, and then you'll start the `lirc` daemon and test that the button codes are being correctly decoded. Once you have installed `lirc`, you should have a `/dev/lirc0` device. You can check by using this:

```
$ ls -l  /dev/lirc*
crw-rw----  1 root root 61, 0 2006-05-18 13:16 /dev/lirc0
```

You might think that since /dev/lirc0 has a zero in it you could have more than one remote control connected to the one system (perhaps /dev/lirc1, /dev/lirc2, and so on). You are correct—you can. You might already have this if you have two TV tuner cards that both came with remote controls.

Tip If you're unsure whether your remote control is sending any signal (for example, the batteries could be flat), you can get a digital camera or camcorder (with an LCD screen) and point it at the remote. Cell phone cameras also work well. The LED that transmits the IR signal from the remote will glow white when it's transmitting. If it doesn't glow, check the batteries. This works because digital cameras also capture a bit of the infrared spectrum of light. This trick doesn't work too well with digital SLRs, though, because most can't display a live preview image.

Using the mode2 Program to Test the Driver

The next step is to use mode2 to test that your remote control is working. The mode2 program talks directly to the hardware device, getting back raw codes. The irw program you will use in a moment talks to the lirc daemon (lircd) and can decode this raw information. If you're using a remote control that came with a TV card, you will likely get similar output, because the hardware on the TV card does the decoding. With more universal IR receivers for remotes, mode2 will get raw pulses (see the same output), while irw will decode these pulses via lircd. So, running mode2 will tell you whether your hardware is hearing anything from your remote, and running irw will tell you whether the lirc daemon is correctly receiving and translating those pulses.

This is example output from running mode2 and pressing some buttons:

```
$ sudo mode2 -d /dev/lirc0
code: 0xc03f30cf
code: 0xc03f30cf
code: 0xc03f30cf
code: 0xc03f30cf
code: 0xc03f08f7
code: 0xc03f08f7
code: 0xc03f08f7
code: 0xc03f08f7
```

In the previous example, we pressed some buttons on the remote, and when we did, several lines popped up on the screen (code followed by a code). This shows you that your remote control is sending an IR signal and the driver is receiving one. Remember, if you have built your own receiver and are not using one from a TV card, you will see raw input of pulse and gaps (shown next). Most TV cards decode the pulses into the codes themselves (like shown previously).

```
$ sudo mode2 -d /dev/lirc0
space 2681403
pulse 9058
space 4415
pulse 633
```

```
space 494
pulse 622
space 510
pulse 608
```

To quit this program, press Ctrl+C.

Configuring the lirc Daemon

The next step is to get the `lirc` daemon understanding what your remote is sending. For remotes that came with TV cards, it should be easy to find a `/etc/lircd.conf` file that somebody has already made and tested. If you have a custom remote and receiver, you can create your own `lircd.conf` using the `irrecord` program.

On the `lirc` website (http://www.lirc.org), there is now a `remotes.tar.bz2` file. It contains all the supported configuration files. Odds are your remote has an existing configuration file, and it's in this archive. For example, we have a Leadtek Winfast TV Deluxe 2000. Inside the `leadtek` directory in the `remotes.tar.bz2` file, you'll find three possible configuration files. (You can browse the `leadtek` directory by double-clicking it and using the Gnome Archive Manager—if you're not running a Gnome environment, you can extract the archive by running `tar xfj remotes.tar.bz2` from a command line.) After looking at each of these configuration files, we decided the RM-0010 configuration was the most appropriate for our remote. We concluded this by finding which configuration file had all the buttons listed that we could see on our remote control. The other configurations might work too (in our case they did), but only for a subset of the buttons.

Once you've found the configuration file, you need to copy it to `/etc/lircd.conf`, and since this is under the `/etc` directory, you'll need superuser privileges to do so. If you copied the file out of the archive to the desktop (by dragging and dropping it from Archive Manager or by clicking Extract), you can issue the following command:

```
$ sudo cp Desktop/lircd.conf.RM-0010 /etc/lircd.conf
```

Now you can start the `lirc` daemon, which will read the `lircd.conf` file and decode the signals from your remote control.

Starting lircd

You can start `lircd` manually like so:

```
$ sudo lircd -d /dev/lirc0
```

You need to specify the device for all the `lirc` processes as the default—`/dev/lirc`—isn't available on more modern versions of the Linux kernel that create the files in `/dev` dynamically. You will be running one of these kernels unless you have specifically gone out of your way not to do so.

Unfortunately, `lircd` starts (or doesn't) silently. If, for example, you forgot to specify the device parameter with `-d /dev/lirc0`, you won't get an error message; `lircd` just won't stay running. You can check whether you have an `lircd` running by checking the output:

```
$ ps -e|grep lirc
  4841 ?        00:00:00 lirc_dev
 14217 ?        00:00:00 lircd
```

Here you can see that we do have lircd running. If we didn't, the output would be more like this:

```
$ ps -e|grep lirc
 4841 ?        00:00:00 lirc_dev
```

If you want to double-check and potentially see some error messages, you can tell lircd not to fork to the background. That is, run in the foreground, not letting you enter in any other commands until you press Ctrl+C to quit it. You can do this by supplying the -n flag (shorthand for --nodaemon) to lircd:

```
$ sudo lircd -n -d /dev/lirc0
```

You will probably want to have the lirc daemon start when your MythTV box starts. You will need to copy the contrib/lirc.debian file from the lirc source tree into /etc/init.d/ (possibly renaming it to just lirc and definitely editing it so the device filename is correct) and run the update-rc.d program like this:

```
$ sudo update-rc.d lirc defaults
```

This is one of those spots that will differ greatly if you're not using Ubuntu; you'll need to check your distribution's administrator guide to determine how to start it (try man chkconfig).

Using irw to Check Your lircd Configuration (/etc/lircd.conf)

The irw program uses the lirc daemon (lircd), which reads the /etc/lircd.conf file and translates these codes into button names. For example:

```
$ irw
00000000c03fc03f 00 FULLSCREEN RM-0010
00000000c03fc03f 01 FULLSCREEN RM-0010
00000000c03fc03f 02 FULLSCREEN RM-0010
00000000c03fc03f 03 FULLSCREEN RM-0010
00000000c03f08f7 00 CH_DOWN RM-0010
00000000c03f08f7 01 CH_DOWN RM-0010
00000000c03f08f7 02 CH_DOWN RM-0010
00000000c03f08f7 03 CH_DOWN RM-0010
```

To quit this program, press Ctrl+C. Now that you have the remote, the driver, and the lirc daemon all working, you can configure the lircrc file for your remote.

Mapping Buttons on the Remote to Functionality

The lircrc file maps buttons on the remote control to functionality in software. This will likely be different for each remote control and for each user's personal preference. By "user," we mean the user that the MythTV frontend runs as. Although you can theoretically have different setups for different logins and different remotes, it can be complicated.

For example, looking at our remote (Figure 3-1), we've decided that the volume, full screen, and channel buttons would be better arranged as the arrow keys and the Enter key for the purposes of MythTV (see Figure 3-1 for the remote controls). The following .lircrc snippet does this:

```
begin
    prog = mythtv
    button = CH_UP
    repeat = 4
    config = Up
end

begin
    prog = mythtv
    button = CH_DOWN
    repeat = 4
    config = Down
end

begin
    prog = mythtv
    button = VOL_UP
    repeat = 4
    config = Right
end

begin
    prog = mythtv
    button = VOL_DOWN
    repeat = 4
    config = Left
end

begin
    prog = mythtv
    button = FULLSCREEN
    repeat = 4
    config = Space
end
```

Configuring lirc for MythTV

You are going to want to map buttons on your remote to functions in MythTV. These don't have to be what is written on the remote. In fact, we've mapped only one button on our remote to the function that's written on the button. This can lead to some confusion for guests picking up the remote, but we've mapped it so that the commonly used functions are easy to access and near each other—the most important being the arrow keys (see the leftmost remote in Figure 3-1 and the previous configuration snippet).

Configuring lirc for Xine

Xine is often used for the playback of some media types in MythTV. With MythTV 0.20, the built-in DVD player is quite good, and a lot of people use it instead of Xine. However, Xine is

still used to play back other videos. Xine has a nice feature where it will generate a template
.lircrc with all the possible commands for Xine. You can append this template to your .lircrc
by doing the following:

```
$ xine -keymap=lirc >> .lircrc
```

> **Note** This is a double greater-than sign; if you're new to Linux, that's very important. It means that the
> output of Xine should be appended to the lircrc file, instead of overwriting it, which is what happens if you
> use only one greater-than sign.

The start of the section appended to the file will look something like this:

```
##
# xine key bindings.
# Automatically generated by xine-ui version 0.99.3.
##

# start playback
begin
        remote = xxxxx
        button = xxxxx
        prog   = xine
        repeat = 0
        config = Play
end

# playback pause toggle
begin
        remote = xxxxx
        button = xxxxx
        prog   = xine
        repeat = 0
        config = Pause
end
```

You now get to edit your .lircrc file and fill in the remote and button parts (possibly the
repeat value as well, depending on your remote) to suit your tastes. You don't need to delete
the remote button mappings that you don't use; you can keep them for future reference.

In our experience, it took a little while to get the mappings just right. Expect to tweak
.lircrc, restart lircd (which is how you reload the lircrc file when you've changed it), check
in Xine, and repeat.

Finding Further Documentation for lirc

The lirc website (http://www.lirc.org/) has pointers to a lot of resources on remote controls.
Some documentation can be a bit out-of-date, so your mileage might vary on how much you

have to tweak the instructions for your setup. However, it is a complete resource that covers most questions. Documentation on hardware compatibility, building your own receiver, and using lirc for sending infrared signals to other devices might prove invaluable for your setup.

Building MythTV

You are now ready to compile MythTV; this takes the source code you have downloaded and turns it into program files that are executable by the computer.

First, run the configure program to get MythTV configured for compilation. configure determines system-specific information; many open source developers use it to create packages that work on a large variety of systems. Here you can set some specific options about how you want MythTV to be built. The simplest method is to run the configure program without any options. This will look something like this:

```
$ cd mythtv-0.20
$ ./configure

# Basic Settings
Compile type     release
Compiler cache   yes
DistCC           no
Install prefix   /usr/local
CPU              x86 (model name        : AMD Sempron(tm) Processor 2600+)
Big Endian       no
MMX enabled      yes

# Input Support
Joystick menu    yes
lirc support     yes
Apple Remote     no
Video4Linux sup. yes
ivtv support     yes
FireWire support no
DVB support      no [/usr/include]
DBox2 support    yes
HDHomeRun sup.   yes
CRC Ip Rec sup.  yes
FreeBox support  yes

# Sound Output Support
OSS support      yes
ALSA support     yes
aRts support     yes
JACK support     no
DTS passthrough  no

# Video Output Support
x11 support      yes
```

```
xrandr support     yes
xv support         yes
XvMC support       no
XvMC VLD support   no
XvMC pro support   no
XvMC OpenGL sup.   no
Mac accel.         no
OpenGL vsync       no
DirectFB           no

# Misc Features
Frontend           yes
Backend            yes

# Bindings
bindings_perl      no
Creating libs/libmyth/mythconfig.h and libs/libmyth/mythconfig.mak
```

If you are going to be using a DVB card (Digital TV) or a remote control, you'll need to specify some options to configure. For initial simplicity, we'll just continue with enabling DVB support by specifying --enable-dvb and --enable-proc-opt (which is recommended by the MythTV configure script when enabling DVB because of the extra CPU required to process digital television). If you're using lirc for remote control, you need to also use --enable-lirc as well. Running configure with these options, the output should look something like this:

```
$ ./configure --enable-dvb --enable-proc-opt --enable-lirc

# Basic Settings
Compile type      release
Compiler cache    yes
DistCC            no
Install prefix    /usr/local
CPU               x86 (model name        : AMD Sempron(tm) Processor 2600+)
Big Endian        no
MMX enabled       yes

# Input Support
Joystick menu     yes
lirc support      yes
Apple Remote      no
Video4Linux sup.  yes
ivtv support      yes
FireWire support  no
DVB support       yes [/usr/include]
DBox2 support     yes
HDHomeRun sup.    yes
CRC Ip Rec sup.   yes
FreeBox support   yes
```

```
# Sound Output Support
OSS support        yes
ALSA support       yes
aRts support       yes
JACK support       no
DTS passthrough    no

# Video Output Support
x11 support        yes
xrandr support     yes
xv support         yes
XvMC support       no
XvMC VLD support   no
XvMC pro support   no
XvMC OpenGL sup.   no
Mac accel.         no
OpenGL vsync       no
DirectFB           no

# Misc Features
Frontend           yes
Backend            yes

# Bindings
bindings_perl      no
Creating libs/libmyth/mythconfig.h and libs/libmyth/mythconfig.mak

WARNING: When using --enable-proc-opt you must include the
         output of ./configure along with any bug report.
```

You are now ready to compile MythTV. You do this by running the make program like so:

```
$ make
```

It used to be necessary to manually run qmake before running make, but the configure program now does this for you. Lots of output will now be printed to the screen. This is building each part of MythTV, which is composed of many source files. How long this takes will vary based on the speed of your machine, the number of processors in your machine, and the speed of your disks. It took around 15 minutes on one of our machines. After it has completed, you can now install MythTV. You do this by executing make install with the appropriate permissions—this means running make install with sudo, like this:

```
$ sudo make install
```

You might be prompted for your password. All the files needed by MythTV will now be copied into directories under /usr/local/.

> **Note** If you get a mysterious error like this:
>
> ```
> distcc[16356] ERROR: compile (null) on localhost failed
> In file included from mythmainwindow.h:8,
>                  from mythmainwindow.cpp:22:
> /usr/include/qt3/qgl.h:79:20: error: GL/gl.h: No such file or directory
> ```
>
> it means you are missing some necessary graphics libraries; therefore, run the following:
>
> ```
> $ sudo apt-get --reinstall install mesa-common-dev
> ```

Configuring MythTV for the First Time

You are now ready to start configuring MythTV. The first decision you will have to make is where you are going to store your recordings. When you configured the disks in your MythTV machine, you should have created a large volume (partition) and set a mount point. We typically create a large /home partition and store everything there, although a lot of people use a file system called /data/mythtv. The default is /mnt/store. You need to create a directory within that partition for the recordings to be stored in too. You can do this via the graphical interface or from the command line. For example:

```
$ mkdir ~/myth-recordings
```

The ~ part means your home directory, so since we're logged in as the myth user, we've just created the /home/myth/myth-recordings directory using mkdir. Next you can run the mythtv-setup program. It's a graphical program that will guide you through setting up some of the basics. Run it with this:

```
$ /usr/local/bin/mythtv-setup
```

> **Tip** You might find that you don't have a mythtv-setup program. On some distributions it is named mythtvsetup.

You'll first be asked a rather simple question—what language you want for MythTV. Choosing a language you understand helps the process (rather a lot actually), as shown in Figure 3-3.

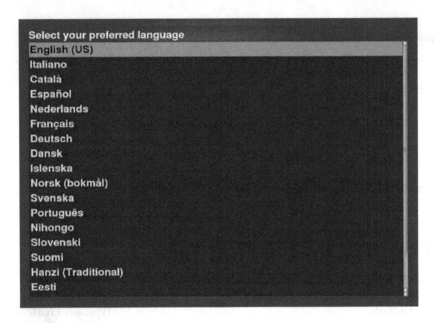

Figure 3-3. *Selecting the language in* mythtv-setup

For the setup process, you can use the keyboard to navigate the options. The important keys are the arrow keys. Up and down cycles through the options, and left and right changes the currently selected option. After you've selected the language, the next setup screen asks you about connecting to the MySQL database (Figure 3-4). Seeing as you're using the default MySQL setup, the defaults here are also fine.

Note In the following section, running the mythtv-setup program, many screens except for the first one (selecting your language) will have Next and Cancel buttons at the bottom. It is possible that these could be pushed off the bottom of your TV set or monitor. If that is the case, you can try changing the settings on your monitor to resize the picture so it fits, or you can adjust the TV-out settings using the relevant tool (such as nvidia-settings). We sometimes recommend using a computer monitor as an interim until you fix any TV output problems.

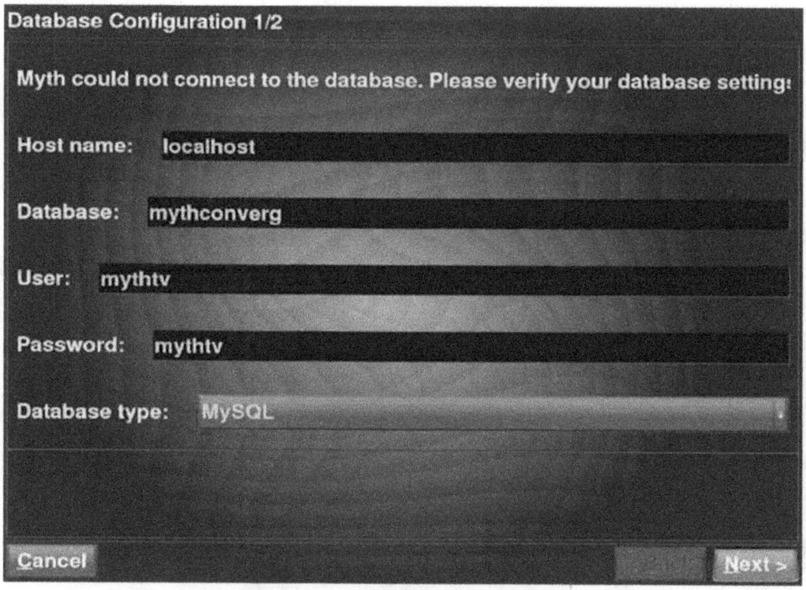

Figure 3-4. *Setting the database connection settings*

Next you configure information about this frontend. A *frontend* is the MythTV component that presents a user interface to the user and displays recordings and videos. This will become more important when you have more than one frontend, which is something we show you how to do in Chapter 8.

The settings for each MythTV frontend (shown in Figure 3-5) are stored by default along with the host name of the frontend. If your host name changes often (for example, if it's retrieved from DHCP, which is common with some Internet providers, although connecting your MythTV box directly to the Internet—without a router in between—is strongly discouraged), you might want to enter an identifier here.

After this screen, you'll be shown the regular mythtv-setup menu, as shown in Figure 3-6. You might notice the setup program "vanish" for a short period of time. This is when it sets up the database with default values and then starts again, with these values. This can cause the theme to change, which is why it looks different in Figure 3-6.

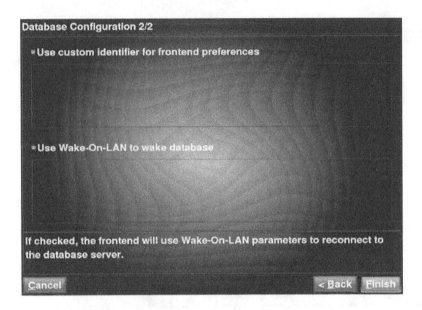

Figure 3-5. *Setting more database parameters*

Figure 3-6. *The main* mythtv-setup *menu*

You can now configure the general settings for MythTV, as shown in Figure 3-7.

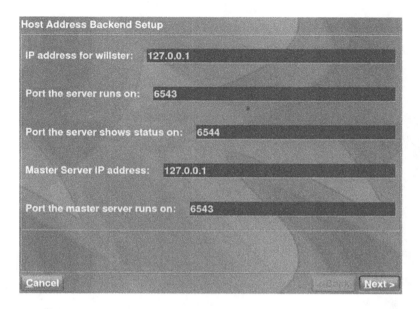

Figure 3-7. *First screen of general settings*

Since we are currently running our backend and our sole frontend on the same machine, the IP address for our host name (in this case, the machine is called *willster*) can be localhost (or 127.0.0.1, the IP address that always means "this machine"). Since we're all on the same machine, the master backend is this machine. The default port numbers are fine. If later you set up multiple MythTV backends, you will need to change the IP address settings to match the IP addresses of the relevant machines.

The next option to configure is where to store recordings. Previously you created the directory, so change the setting to the directory you created before, such as /home/myth/myth-recordings, as shown in Figure 3-8.

Note that in Figure 3-8 we're showing the default path that MythTV wants to use for your recordings. You should change this to wherever you set up the directory if it's not at the default location. The checkbox tells the backend to use a special method for deleting large files, in case you chose not to use a file system (such as XFS) that handles the deletion of large files gracefully. If you need this option and don't set it, then deleting recorded programs might sometimes appear to freeze your system for a few seconds.

Next, we'll show how to set some global settings for the backend, such as the TV format and frequency table, as shown in Figure 3-9.

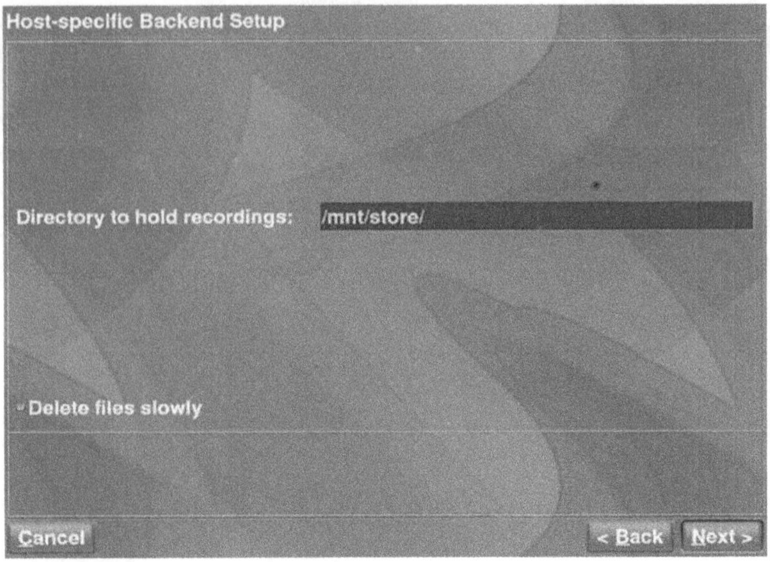

Figure 3-8. *Selecting where to store recordings*

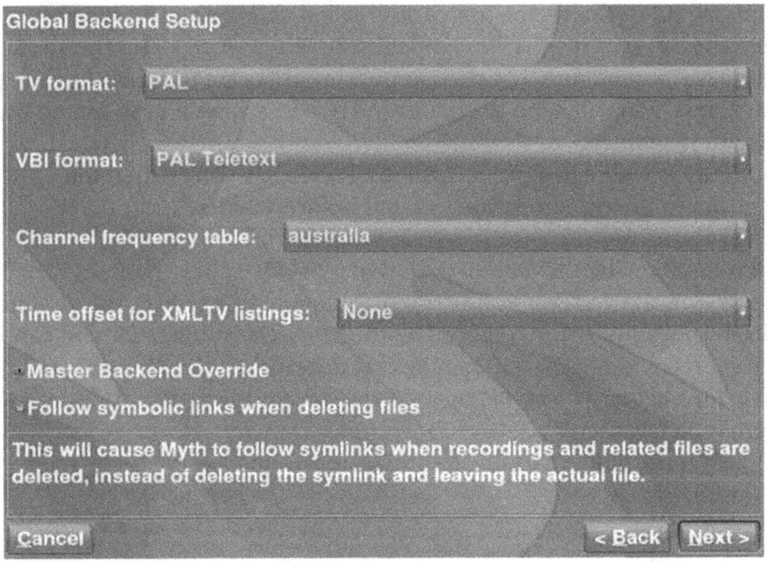

Figure 3-9. *Selecting TV format in* mythtv-setup

The TV format depends on where you live; for a more complete discussion of the various format options, see Chapter 1. Here you can see we've set the TV format to PAL (which is used in Australia), the VBI (which stands for *vertical blanking interval*) format to PAL Teletext, and the channel frequency table to Australia. The VBI format is the format of closed-captioning data. In some areas, you might need to change the Time Offset for XMLTV Listings settings, although this can largely depend on where you get your TV guide data from; we cover guide data in Chapter 4.

The next configuration screen (shown in Figure 3-10) defines how MythTV handles the event information table (EIT), which is the program listing that is sometimes broadcast along with the television signal with digital broadcasts. This is different from the guide data you feed MythTV that's discussed in Chapter 4. Often, the EIT data isn't for an extended period of time—perhaps only for the next day. However, in some areas, such as Australia, the free-to-air television stations just provide the bare minimum of information: the current and next programs. To add to the fun, sometimes the EIT information is not very descriptive or has different names for programs than your guide data, possibly causing you to miss recordings. Even with all these caveats, it can still be useful for last-minute changes.

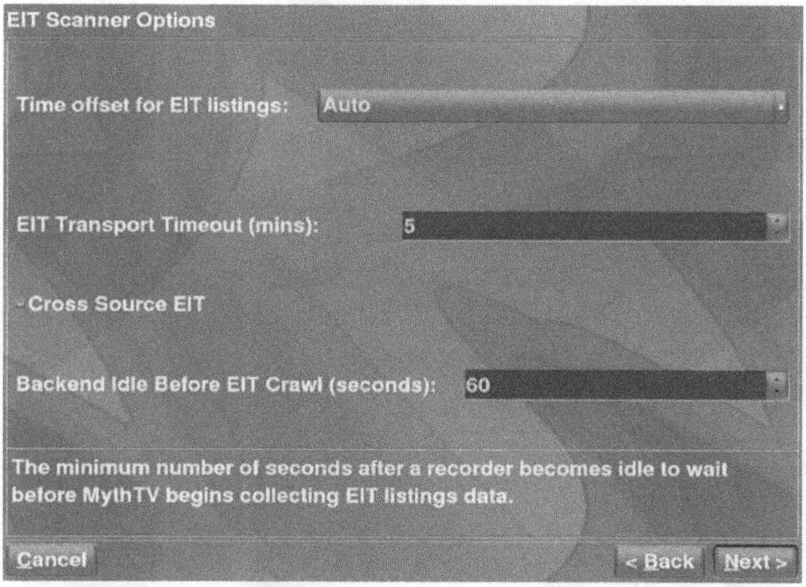

EIT Scanner Options

Time offset for EIT listings: Auto

EIT Transport Timeout (mins):   5

Cross Source EIT

Backend Idle Before EIT Crawl (seconds):   60

The minimum number of seconds after a recorder becomes idle to wait before MythTV begins collecting EIT listings data.

Cancel   < Back   Next >

Figure 3-10. *EIT settings*

How frequently MythTV checks for EIT data (and how long it spends waiting for it) depends on these configuration options. The EIT Transport Timeout setting is how long MythTV will spend waiting on *each* channel for EIT data. If there are a lot of channels, this could mean a very long time before MythTV finds the last-minute schedule change and acts accordingly. You might choose to tune this option once you have some real-world experience in your area.

For the rest of the general settings (Figures 3-11, 3-12, 3-13, and 3-14), the defaults are probably fine; some are applicable only if you have multiple backends. You might want to tweak some of the settings (for example, when jobs can run and how many at once) depending on the processing power of your system. Some people set up various machines to automatically power on and off using various methods (including WakeOnLan) in an effort to save power or make their systems quieter. You can configure some of these settings here, although we don't go into detail about them in this book.

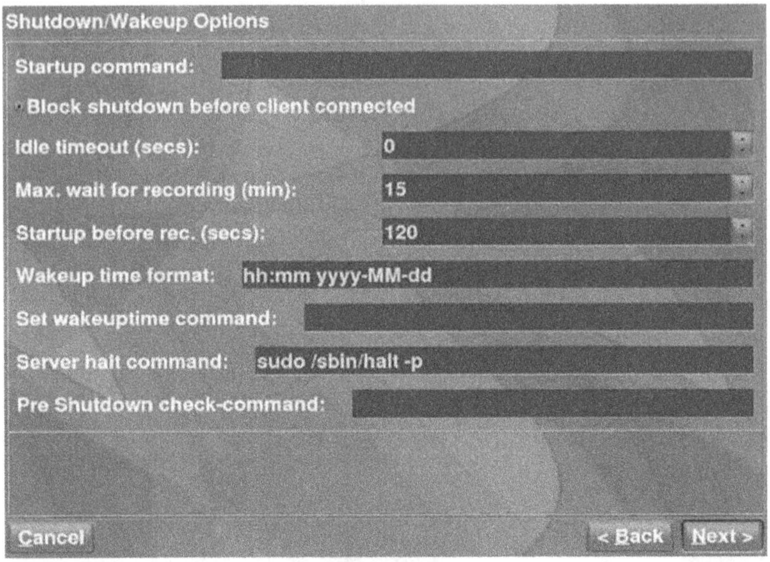

Figure 3-11. *Shutdown/Wakeup Options screen*

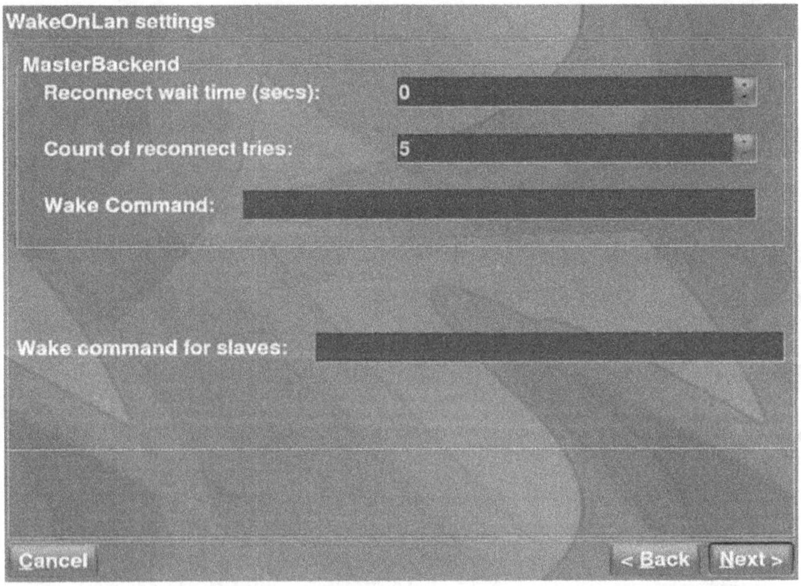

Figure 3-12. *WakeOnLan Settings screen*

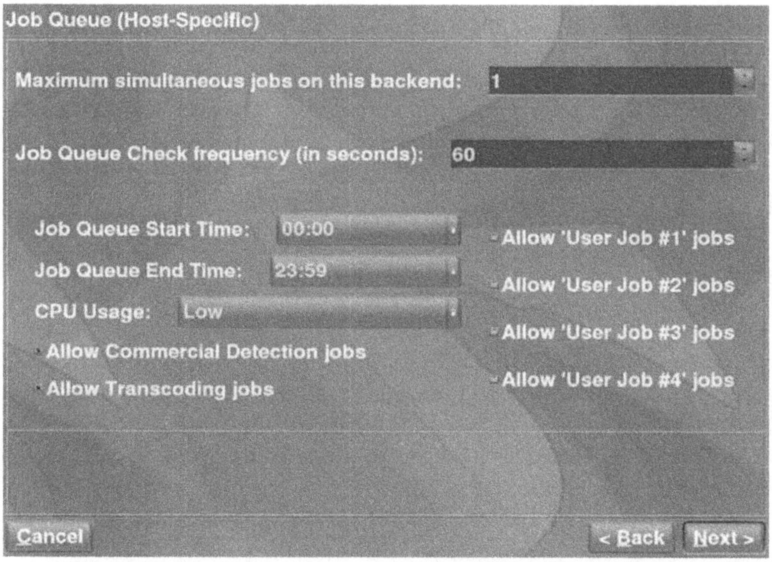

Figure 3-13. *Job Queue (Host-Specific) screen*

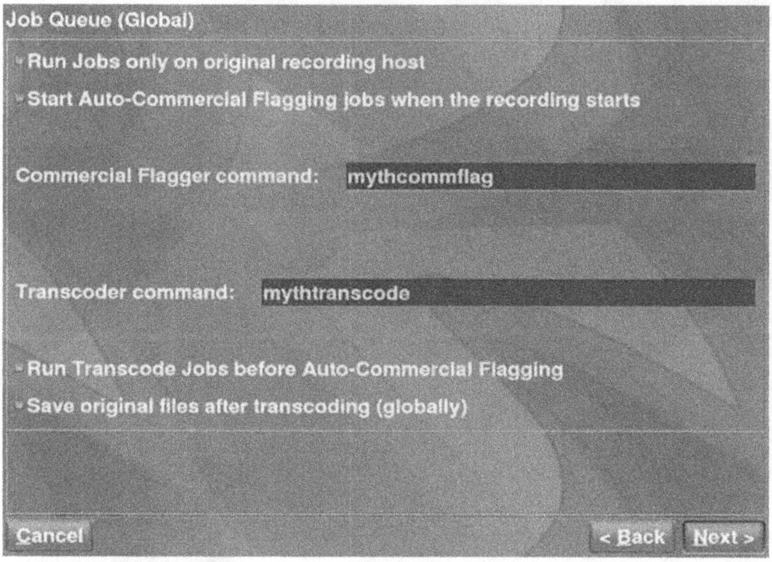

Figure 3-14. *Job Queue (Global) screen*

The configuration options shown in Figure 3-14 are useful if you have multiple backends and some of them are over a slower network such as wireless. With multiple backends, MythTV might decide that a recording made on one can be transcoded or that commercial flagging can be performed on a different machine (the idea is to balance CPU usage across all backends). The Run Jobs Only on Original Recording Host option prevents this approach, but we've found this might cause problems on slower network links (such as wireless), especially when also running frontends over the same wireless network. There are also options for setting when jobs start, which you might decide to change if you have extra CPU power.

When initially configuring your MythTV system, you likely do not know how you should set many of these options. In this case, the defaults are often the best. However, after a period of use, you might discover some behavior that you want to change. You can always run `mythtv-setup` again and change these settings, and it might be a good idea to revisit these settings after you've been running MythTV for a month or two.

After completing the general setup, you are now ready to set up your capture cards.

Setting Up the Capture Cards

As we've previously discussed, it's possible to use multiple TV tuner cards with MythTV. It's also possible to use both analog and digital cards. In previous versions, setting up analog and digital cards was quite different, possibly involving jumping through some nonobvious hoops. The good news is that now it's quite easy for many common cards. Selecting Capture Cards from the `mythtv-setup` menu, as shown in Figure 3-15, will take you to the list of current capture cards. Since you are adding your first capture card, this list is blank. If you have several cards or are changing some settings, you'll notice your existing cards in the list.

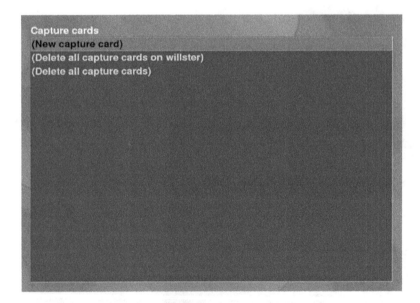

Figure 3-15. *Setting up capture cards in* `mythtv-setup`

You'll need to configure each capture card you have separately. Configuration differs slightly depending on the type of capture card. MythTV supports the following types of input devices:

- Analog Video4Linux capture card

- MJPEG capture card (Matrox G200, DC10)

- MPEG-2 encoder card (Hauppauge PVR-x50, PVR-500)

- DVB DTV capture card (v3.*x*)

- pcHDTV DTV capture card (with Video4Linux drivers)

- FireWire cable box (including the DCT-6200, SA3250HD, and SA4200HD, along with support for "other" boxes)

- USB MPEG-4 encoder box (Plextor ConvertX, and so on)

- Dbox2 TCP/IP cable box

- HDHomeRun DTV tuner box

- RCR IP Network Recorder

- Freebox Network Recorder

The most common are analog Video4Linux cards, DVB DTV capture cards, and Hauppauge PVR MPEG-2 cards. Configuring the other types of cards is similar (or even simpler than these).

For an analog card, if you have just one, the device will be /dev/video0 (the default). If you have more than one card, you'll be able to select several devices (all starting with /dev/video and ending in a number). Note that webcams can also show up as /dev/video devices. If you have multiple devices, the Probed Info line might help. The audio interface might be your sound card (probably /dev/dsp) if there is a pass-through cable connecting a sound output from the capture card to the sound input of your sound card. Some analog cards have a "sound card" (usually only supporting sound coming in from the TV station) that's usable. If yours does, it's best to use this. It will likely show up as another audio device such as /dev/dsp1. Some experimentation might be required depending on the number of possible combinations on your system. If you are using a card that supports MPEG digital data, such as a Hauppauge or DVB card, then the audio is usually embedded in the video stream; this is almost a requirement for using more than one or two tuners in a backend. Figure 3-16 shows adding an analog Video4Linux capture card.

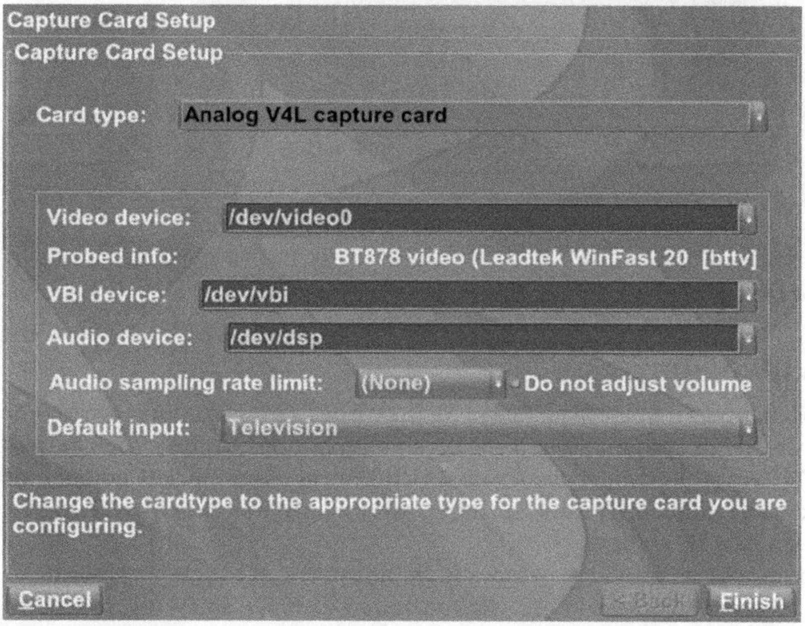

Figure 3-16. *Adding an analog Video4Linux capture card*

When setting up a digital (DVB) card, the defaults should be OK, but make sure there is a "frontend ID" (it will display the name of the chipset that's on your DVB card—not the manufacturer of the card). See Figure 3-17. If you have multiple digital capture cards, for each card you add, the DVB card number should be different. The signal and tuning timeouts should be decent values. If your digital card is connected to digital satellite equipment that supports DiSEqC, you can configure MythTV to talk to your satellite equipment correctly.

The Recording Options button's screen allows you to configure a few more obscure options, the most notable being Open DVB Card on Demand, which, if selected, frees the DVB card to be used by other applications when MythTV is not recording. However, this means that if another application is using the DVB card when MythTV wants to use it, MythTV will be unable to use the card. Since the scheduler cannot predict when the card might be in use by another application, you should probably avoid this option unless you have a good reason to enable it.

In previous versions of MythTV (prior to 0.20), after you've set up the capture card, MythTV would scan for TV channels onto which it could lock. This step is now part of setting up video sources (covered later in this chapter).

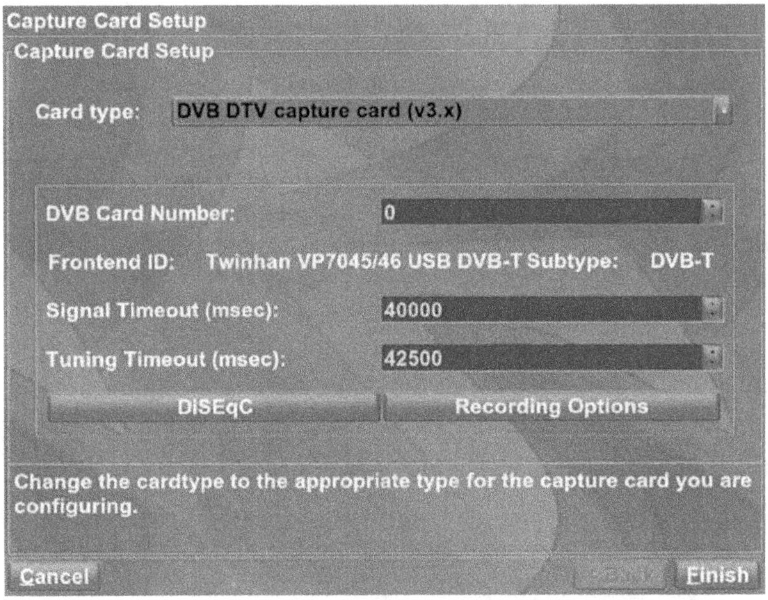

Figure 3-17. *Setting up a DVB card*

Setting Up the Video Sources

A *video source* is a set of channels that are available to MythTV from a program source, whether satellite, cable, or an over-the-air antenna. You might have multiple video sources, such as free-to-air TV and pay-for TV. Here is also where you set up which TV programming data grabber to use. If you're lucky, the default grabber will work OK; otherwise, in some areas (such as Australia), you might have to install a new TV grabber.

To configure video sources, select Video Sources from the mythtv-setup menu. You will see a listing of your existing video sources along with the option to create a new one or delete the existing one (see Figure 3-18).

Figure 3-18. *Adding video sources*

You can name each video source so you can remember what it is. In Figure 3-19 we've named the video source *freetoair*, selected the appropriate grabber for Australia, and selected the channel frequency table. We've also enabled EIT to supplement the guide data with listing information broadcast over the air with the DVB signal.

Video source setup

Video source name: freetoair

XMLTV listings grabber: Australia

tv_grab_au: Configuration will run in the terminal window

Perform EIT Scan

Channel frequency table: australia

If this is enabled the data in this source will be updated with listing data provided by the channels themselves 'over-the-air'.

Cancel   Finish

Figure 3-19. *Setting up the video source with configuration for Australia*

Once you've set these settings, MythTV will run the configuration program for your grabber. Depending on the grabber, this might take the settings from the Video Sources screen (for example, the North American grabber), or it might require you to switch to the terminal window that you ran `mythtv-setup` from and answer some configuration questions. Use Alt+Tab to switch back to the terminal window to answer these configuration questions. When you've finished configuring the grabber, you can Alt+Tab back to `mythtv-setup`. If you haven't yet installed a grabber, you might be presented with an error message. It's safe to continue; you just won't have any guide data yet.

Configuring Input Connections

The Input Connections screen (in `mythtv-setup`) links input connections from capture cards to video sources. You can link multiple capture cards to the same video source. You'll see a list of all the connections on all your currently configured capture cards (see Figure 3-20). Selecting the connection presents you with the screen in Figure 3-21. From here you can (optionally) give this a name and select the video source (configured previously).

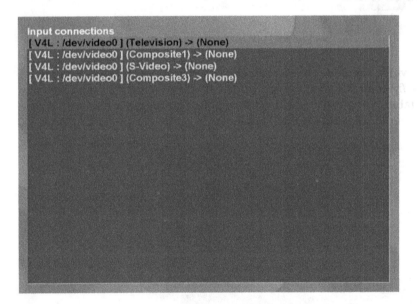

Figure 3-20. *Viewing the input connections listing*

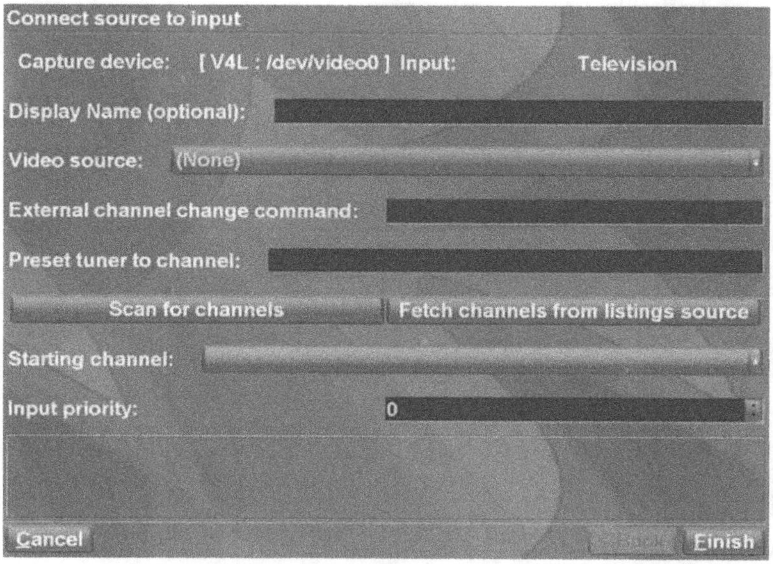

Figure 3-21. *Configuring input connections for a capture card*

Some setups might require the execution of an external command to change the channel. A common example is using an IR blaster to let MythTV change the channel on a cable TV set-top box. You can also now use the Scan for Channels or Fetch Channels from Listings Source option to set up all the channels you can receive. In MythTV 0.20, it's now easy to use the scanning feature for both analog and digital capture cards; this is a good method to get things configured quickly and correctly. Fetching channels from the listing source is more common for cable TV setups and any setup using an external box to change channels.

Figure 3-22 shows how to configure the scanning for channels. You can choose which capture card to use to scan for the video source, as well as the type of scan. Usually the defaults are fine.

You can also set an input priority for the capture source. For example, if you have two cards, such as an old analog one and a new digital one, you might want the digital card to be used instead of the analog one. To do this, give the digital card a higher input priority.

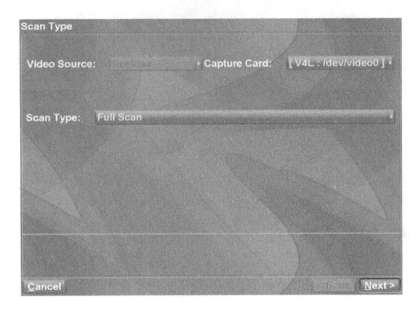

Figure 3-22. *Configuring scanning for channels*

Setting Up the TV Channels

The Channel Editor lets you set up each TV channel you can receive, linking it to guide data and specifying whether it has commercials (if it doesn't, there's no point in running commercial detection jobs on it). Select the Channel Editor from the mythtv-setup menu. If you have a DVB card and have scanned for channels (and found some), they'll be listed (see Figure 3-23).

Figure 3-23. *Viewing the Channel Editor list*

After selecting a channel to configure, you can set the name of the channel to be shown in the GUI. The channel number is what number the TV card should tune to, and the call sign is a short name for the channel (see Figure 3-24 for an example). Other important fields to set are as follows:

Icon: You can specify a path to an image (some TV guide grabbers also grab an image) that will be the icon for that station. When you're changing channels, this icon is displayed.

XMLTV ID: This is the ID from the guide data grabber for the channel. Some grabbers let you specify the ID for each channel it grabs. You might have to examine the output of the grabber (for example, by running tv_grab_au|less).

Commercial Free: If the TV station doesn't have commercials, check this box.

Visible: You might want to hide some channels from the user interface—these might be channels that just broadcast programming guides.

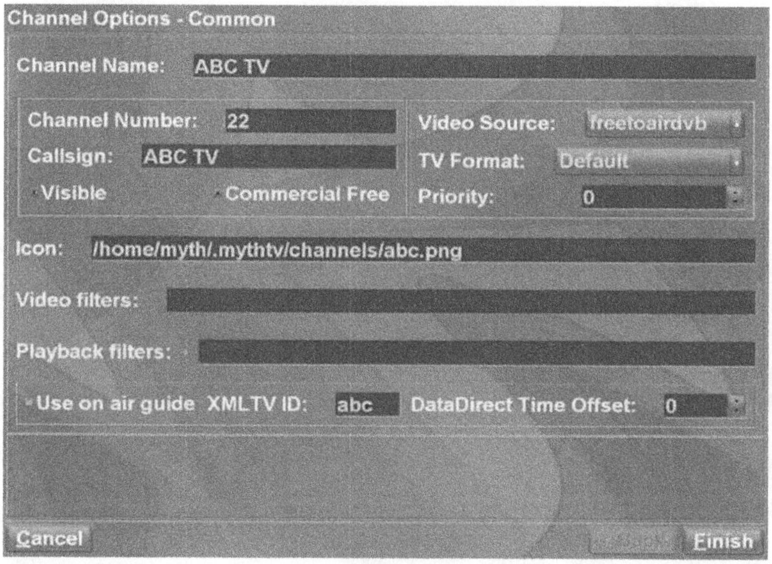

Figure 3-24. *Channel options*

Starting the Backend

Now that you have most options set up, you can start the backend, then start the frontend, and finally do a quick test to make sure things are working. At the moment, we'll show how to start the backend manually. So as the myth user, open a terminal window, and start the backend:

```
$ mythbackend
```

You should see some log messages printed about connecting to the database, the version of the backend, and perhaps some output from the regular jobs the backend performs (such as Running Housekeeping). To stop the backend, you can press Ctrl+C. Before doing so, you might want to skip ahead slightly to run the frontend and play with your MythTV setup. However, be sure to return here to set it up so that the backend starts automatically.

You will probably want to install an init script so that the MythTV backend will start every time you start your computer, the logs are written to a log file (stored in /var/log), and it shuts down nicely when you shut down the computer running the backend. We use the following script as /etc/init.d/mythbackend:

```
#! /bin/sh
#
# mythbackend
#

PATH=/usr/local/sbin:/usr/local/bin:/sbin:/bin:/usr/sbin:/usr/bin

BACKEND=/usr/local/bin/mythbackend
BACKEND_NAME=mythbackend

MYTHUSER="myth"

test -x $BACKEND || exit 0

set -e

case "$1" in
  start)
        echo -n "Starting MythTV: $BACKEND_NAME"
        su ${MYTHUSER} -c "$BACKEND -v all -d -l ${MBE_LOGFILE}" &
        echo "."
        ;;
  stop)
        echo -n "Stopping $DESC: $NAME "
        killall $BACKEND
        echo "."
        ;;
  *)
        echo "Usage: $0 {start|stop}" >&2
        exit 1
        ;;
esac

exit 0
```

To easily enter this script, you can copy and paste it from http://www.mythtvbook.com/ using this command to launch the editor:

```
$ sudo gedit /etc/init.d/mythbackend
```

You will also need to change the permissions on the file to make it executable:

```
$ sudo chmod +x /etc/init.d/mythbackend
```

To enable this new script so that the backend automatically starts and stops, run the following:

```
$ sudo update-rc.d mythbackend defaults
```

This is one of those spots that will differ greatly if you're not using Ubuntu; you'll need to check your distribution's administrator guide to determine how to start it (try man chkconfig). To rotate the log files so that they don't continue to grow forever and use up all your disk space, use the following configuration file for the logrotate program (which comes preinstalled with the operating system). We ran the following command:

```
$ sudo gedit /etc/logrotate.d/mythtv
```

This uses the program gedit (a text editor) as the superuser to open the new file /etc/logrotate.d/mythtv. We then typed the following in, saved, and quit:

```
/var/log/mythbackend.log {
        copytruncate
        daily
        size 10M
        missingok
        rotate 7
        compress
        notifempty
}
```

Starting the Frontend

Now that the backend is running, you can start the MythTV frontend, which is the graphical interface to MythTV. From a terminal window, you can start the frontend just by typing this:

```
$ mythfrontend
```

Selecting Watch TV should present you with a tuned TV channel (and sound). The up and down arrow keys should change the channel. You will also be able to pause (by pressing P) and rewind (or fast-forward assuming you've paused to let the television broadcast get ahead of you) with the left and right arrows, respectively.

If you have an analog card and you still hear sound when the TV is paused, you might need to mute the sound input channel. To do this, quit the frontend (press Escape to quit) and then open the Volume Control applet by right-clicking the speaker icon on the top right of the screen and selecting Open Volume Control. On the Capture tab, each input source is listed. Click the icon of a speaker to mute the output.

Automatically Starting the Frontend

You will probably want to start the frontend automatically when you log in. To do this, open System ➤ Preferences ➤ Sessions, and select the Startup Programs tab. Here you can add mythfrontend to the list of programs to start when you log in. You can test that it works by logging out and then logging back in.

Conclusion

In this chapter, we showed how to configure a remote control, how to install MythTV, and how to set the backend and frontend to start when you turn on the computer. We also showed how to configure TV channels to allow you to watch Live TV, pause it, and rewind it. In the next chapters, we'll cover setting up guide data as well as basic television recording.

CHAPTER 4

■ ■ ■

Recording TV

Now that you have the hardware configured for your MythTV system and have the software installed, it's time to start recording some TV. We have split the description of the TV-recording functionality in MythTV across this chapter and Chapter 5. This chapter will cover the basic aspects to get you up and running. Chapter 5 covers advanced topics such as how to detect commercials, how to automatically skip commercials, how to set the autoexpiry of video, and how to set the video playback options and transcoding. You'll also learn more about editing recordings (using cut points) in Chapter 6, as well as lots of other MythTV functionality.

The first step to recording TV is to get guide data for your TV providers; this is by far the most complicated bit of the work covered in this chapter, so it consumes most of the space. This is not so much because the task is complicated but because there are so many different providers and ways to get the data exist. Once you have guide data, you can schedule some recordings. We'll show you how to browse through the on-screen guide and select shows to record. More important, we'll then show you how to search for your favorite shows and set them up to be recorded. Finally for this chapter, we'll cover how to resolve conflicts between shows.

This is an important chapter, because although MythTV can perform manual recordings, it's much more useful with good guide data. So, work through the nitty-gritty details, because the outcome is worthwhile.

Getting Guide Data

PVR systems are only as good as their guide data. Obviously, if the PVR doesn't know that your favorite show is coming up, it can't record it. Worse than that, though, if the PVR doesn't have enough guide data going into the future, then it is much less likely to be able to resolve conflicts between shows that you want recorded. That's because PVRs resolve conflicts by looking at future showings of programs and trying to record one of the programs later.

This makes your Internet connectivity important to the operation of your PVR. The PVR will be downloading the guide data from the Internet and therefore needs to be able to get to the network when it runs the update job. If the Internet connection is consistently not available, then you will find that you start to miss out on recordings.

Finally, if the guide data doesn't include elements such as the names of episodes, then the PVR will record the same shows over and over, which not only is annoying but also takes up disk space and valuable tuner time. It is frustrating to miss out on a show you haven't seen just because the PVR was recording a repeat episode of something else.

This section discusses guide data in general, then the specifics of guide data for North America, and then the specifics of guide data for the rest of the world. We're breaking the discussion up like this because the U.S. guide data situation is quite different from the rest of the world. You should read only the sections relevant to your country to end up with a working guide data setup.

Guide Data in General

MythTV's backend stores the PVR's guide data in the MySQL database you set up in Chapter 3. This database also stores your scheduled recordings and the shows you have already watched that therefore shouldn't be recorded again.

MythTV usually gets this guide data through a system called XMLTV (you can find its site at `http://membled.com/work/apps/xmltv/`), which runs customized grabbers for your particular country. Grabbers are generally based on screen scraping, although commercial options are available in some countries. If you find that an XMLTV grabber doesn't currently support your country, then it is also possible to write your own. We discuss XMLTV and grabbers more in the "Guide Data for the Rest of the World" section, which is about guide data for countries other than North America.

SCREEN SCRAPING

Screen scraping is the technique for extracting information from another person's user interface so you can use it in your system. The term originated with mainframe applications where you couldn't change the output format of an existing application but needed some of its data in a new application. Instead, you'd parse the screen output of the first application to find the information you needed.

With web applications, the purpose has changed. Screen scraping is now a technique commonly used to extract information from websites not under your control so that you can use that information in a way that the website's owner either does not support or does not condone.

An example is TV program guide information sites, which are scraped in many countries to produce the XMLTV guide data needed to make MythTV run well. In some countries, the TV guide site owners are upset about this occurring—often because they sell access to that data commercially in an easy-to-use format.

One vulnerability of this technique is that when the user interface changes, the scraper often stops working. One common technique employed by website owners is to therefore change the user interface for their site regularly.

For more information about screen scraping, see the excellent Wikipedia page at `http://en.wikipedia.org/wiki/Screen_scraping`.

Guide Data for North America

Zap2it (http://www.zap2it.com) kindly provides free guide data (well, for only the cost of a quarterly survey) for both the United States and Canada, and it is because of this generosity that North America gets separate treatment in this chapter. Specifically, getting guide data to work for North America is significantly easier than it is for the rest of the world. It's faster because the grabber doesn't have to download many web pages to scrape the guide information, it's more complete that what is often available elsewhere, and the guide data is generally more reliable.

Note XMLTV isn't needed in North America. In other words, the North American guide data from Zap2it doesn't need XMLTV to be installed. This is because the guide data grabber is built into MythTV. You do need a version of wget, which is at least 1.9.1, though.

The first step to using the Zap2it guide data is to sign up for the service with the Zap2it Labs website. Go to Zap2it Labs at http://labs.zap2it.com/, and you will see a page like the one shown in Figure 4-1.

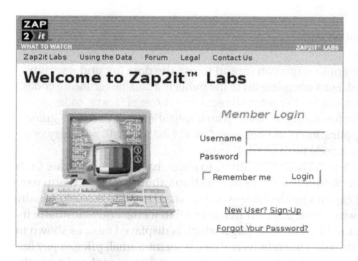

Figure 4-1. *The Zap2it Labs login page*

Click the New User? Sign-Up link to be taken to a license agreement. You'll need to read the entire agreement and then accept it to be able to use the Zap2it guide data. If you don't like the agreement, then you will find getting guide data for North America to be a lot harder, because the XMLTV grabber for North America is no longer supported. If you accept the license agreement, then click the Accept button. Next, you are prompted for a username, a password (twice to be sure), an email address to be contacted at, and a certificate code, as shown in Figure 4-2.

Figure 4-2. *The Zap2it Labs registration page*

Certificate codes identify the application you intend to use the data for, and a variety of certificate codes have been issued. For a complete list of the publicly available certificate codes, check out `http://docs.tms.tribune.com/tech/tmsdatadirect/zap2it/certificate_codes.html`. It is interesting to note that if you're intending to develop an application that needs TV guide data for North America and the application is noncommercial, then Zap2it will seriously consider giving you your own certificate code for that application.

The certificate code for MythTV is ZIYN-DQZO-SBUT, so enter that in the Certificate Code field. Finally, the form asks you to complete a series of questions about your television use. This is the cost you pay for using Zap2it's guide data—you are asked to fill in a questionnaire every three months. You can see when your account needs to fill in its next questionnaire by looking at the Subscription Expires field on the web page, which is displayed next, as shown in Figure 4-3. Given that these questionnaires are relatively short, it seems a small price to pay for free high-quality guide data. In fact, several of the questionnaires we have completed recently have been empty. Don't worry too much about remembering the expiry date, because the MythTV user interface will let you know when your subscription expires. Zap2it also sends email reminders to your registered email address.

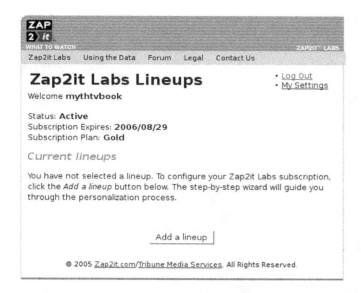

Figure 4-3. *The MythTV user interface will let you know when your subscription expires.*

Now you're ready to add a lineup to our Zap2it subscription. A *lineup* is a series of channels that you receive. Click Add a Lineup. Lineups are arranged by ZIP code, so enter your ZIP code in the field. For this example, we'll use the ZIP code shown in Figure 4-4.

Figure 4-4. *Lineups are organized by ZIP code.*

The Lineup Wizard will then ask you to select how your television is delivered. Options available for the example ZIP code include cable, digital cable, satellite, and local broadcast. You can select only one delivery mechanism per lineup, but you can have more than one lineup if desired—this is one way to handle having a MythTV box that can tune both cable and satellite, for example. We selected Cable here and clicked Next; a list of the cable TV providers for the area then appears. Select your cable TV provider, and click Next. You will see a list of the channels offered in your area by your provider. Note that some cable companies offer multiple lineups in an area, either for regional differences or for digital vs. analog cable; make sure to choose the one that reflects the channels you actually have. Select the channels you receive, and click Finish. You can now add more lineups if desired.

Once you've set up your lineups, you are ready to run the MythTV setup program and let it know your Zap2it username and password. You do this by running mythtv-setup, which will give you a user interface that looks something like Figure 4-5 (the exact screen will vary depending on the theme you are currently using).

1. General

2. Capture cards

3. Video sources

4. Input connections

5. Channel Editor

MythTV

Figure 4-5. *The first screen of the setup wizard*

Tip You might find that you don't have a mythtv-setup program. That's probably because on some distributions it is named mythtvsetup instead.

Use the arrow keys, the Enter key, and the Escape key to navigate the user interface unless you have a remote control configured. If you have a remote control, then the arrow keys, the Select key, and the Back key on the remote should behave as you expect them to behave. Select Video Sources from the menu, and then select New Video Source, as shown in Figure 4-6.

Once you've selected the New Video Source option, you'll see the setup screen for the guide data source. In general, you select your country, which should be North America (DataDirect) for Zap2it guide data. What should you use for a name for the video source? Well, think "program source" rather than "tuner card"—if you have three tuners all tuning one analog cable source, they'll all use *one* video source, probably called something like Your Provider Analog Cable. Figure 4-7 shows our screen after we have configured it.

Figure 4-6. *An empty list of video sources*

Figure 4-7. *The video source setup wizard*

Note that your password appears in clear text on this screen. In other words, our password isn't really composed only of the letter *x*. Note the section of the screen at the bottom, which displays help information for each item in the wizard. Once the video source has been configured, it will appear in the list of available sources, as shown in Figure 4-8.

Figure 4-8. *After the creation of our new video source*

Finally, you need to run `mythfilldatabase` to ensure that there is guide data. You can read more about how to do that in the "Forcing a Guide Update" section.

Other Commercial Providers

Other commercial guide data providers do exist. For example, IceTV (`http://www.icetv.com.au`), which offers guide data for Australian TV channels, doesn't rely on screen scraping to obtain its guide data. It even offers a PDF guide on how to set up its guide data with MythTV, which at the time of this writing is located at `http://www.icetv.com.au/support/howto/howtoiceguide4mythtv.pdf`.

Note `tvguide.org.au` provides free data for Australia. This guide data is arguably less reliable than the IceTV guide data, because volunteers enter the information. It is possible that some days it will have no guide data at all if no volunteer has entered it. In that case, you can donate some time to enter the guide data yourself if you want.

We won't repeat the information from the IceTV PDF here. If you're Linux savvy, it should suffice to say that you need to install XMLTV, download the custom XMLTV grabber, install it with the other grabbers, symlink it to `tv_grab_au`, and then run `mythtv-setup` to configure MythTV. If you're not, grab the PDF file (from the previously mentioned URL), and read the details. We provide a more generic description of how to configure XMLTV grabbers for MythTV in the following section, which should also help if you've never done this before. You can find more information about setting up XMLTV in the next section.

Guide Data for the Rest of the World

As mentioned earlier in this chapter, the more generic approach to getting guide data for MythTV is to use XMLTV to download guide data for your country. XMLTV is a package that produces TV guide data in XML format for various countries. MythTV can then import these XML files and use them to populate its own internal guide data format. XMLTV works by running a custom "grabber" for your country. This grabber collects guide data from the sources appropriate to your country, which might be a commercial source, a free data source, or a screen-scraping website as previously described.

Setting up XMLTV with MythTV is pretty easy, because there are only a couple of gotchas, so let's give it a go. First, you need to have XMLTV installed. Then, if you're using Ubuntu, just use apt-get from a command line:

```
$ sudo apt-get install xmltv
```

You'll be prompted for your password, and apt-get will ask you whether you're sure you want to install XMLTV and its dependencies. Say yes. If you're using a different version of Linux, then you'll need to install the XMLTV package for your distribution or compile it from source. Installing the XMLTV package will also install a selection of grabbers for various countries. These grabbers are actually stored in the xmltv-util package. At the time of writing, the following grabbers are included in this package with Ubuntu Dapper Drake:

tv_grab_au	tv_grab_jp
tv_grab_be	tv_grab_na_dd
tv_grab_br	tv_grab_na_icon
tv_grab_ch	tv_grab_nl
tv_grab_de_tvtoday	tv_grab_nl_wolf
tv_grab_dk	tv_grab_no
tv_grab_ee	tv_grab_re
tv_grab_es	tv_grab_se
tv_grab_fi	tv_grab_se_swedb
tv_grab_fr	tv_grab_uk_bleb
tv_grab_huro	tv_grab_uk_rt
tv_grab_is	tv_grab_za
tv_grab_it	

If the grabber you need is not in this list, then you'll need to find it somewhere on the Internet and install it manually. Installation is as simple as copying the grabber file into /usr/bin. Now you're ready to run mythtv-setup, so make sure you do that from a terminal window (as opposed to, say, using Alt+F2 or Start ➤ Run from a KDE desktop), because you're going to need to be able to see standard output from the program later. You'll see a screen similar to Figure 4-9.

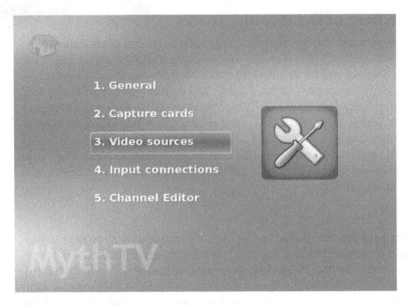

Figure 4-9. *The first screen of the setup wizard*

Note On some distributions, `mythtv-setup` is named `mythtvsetup`.

Use the arrow keys, the Enter key, and the Escape key to navigate the user interface unless you have a remote control configured. If you have a remote control, then the arrow keys, the Select key, and the Back key on the remote should behave as you expect them to behave. Select Video Sources from the menu, and then select New Video Source, as shown in Figure 4-10.

Figure 4-10. *An empty list of video sources*

Once you've selected the New Video Source option, you'll see the setup screen for the guide data source. Figure 4-11 shows an example of a U.K. TV guide being set up.

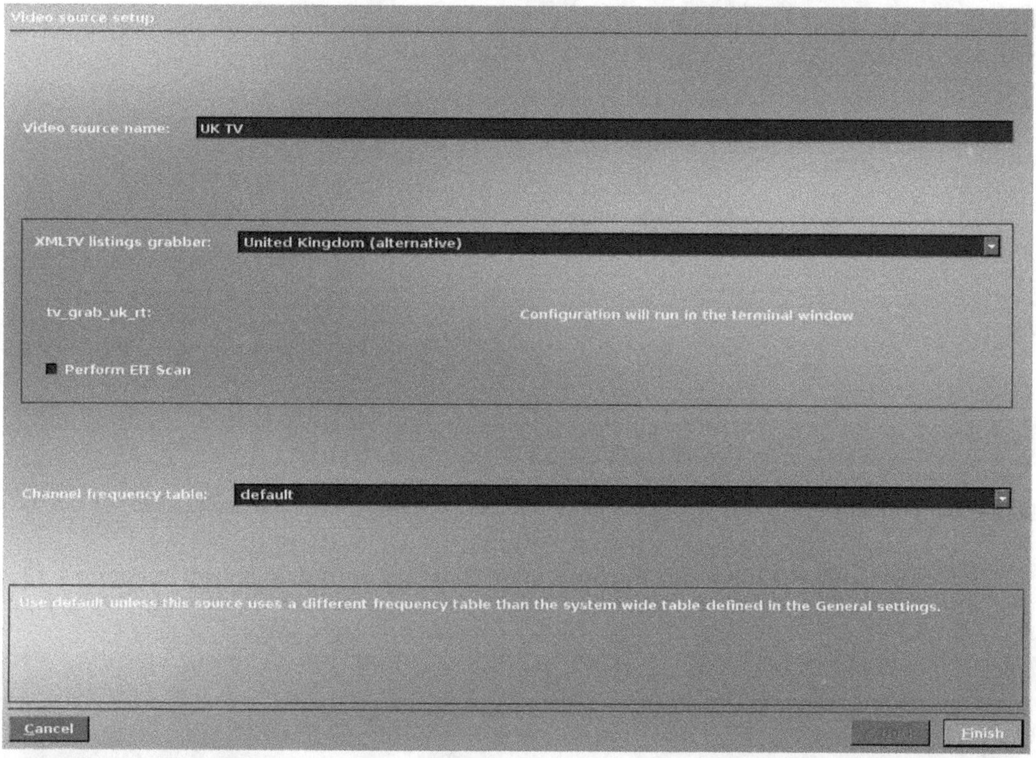

Figure 4-11. *Setting up a U.K. guide data source*

When you hit Finish, then the progress bar will stall at 50 percent for guide data sources that aren't for the Zap2it North American guide data, as shown in Figure 4-12. It's not a crash; it's just mythtv-setup waiting for you to configure the XMLTV grabber on the terminal you started mythtv-setup from (remember we said to start it in a "real" terminal window?).

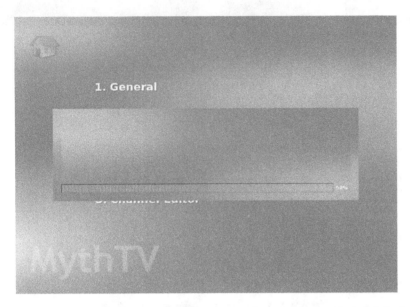

Figure 4-12. *The setup progress bar gets stuck at 50 percent, but this isn't a crash.*

You'll need to hit Alt+Tab to see the prompt on standard output, as shown in Figure 4-13.

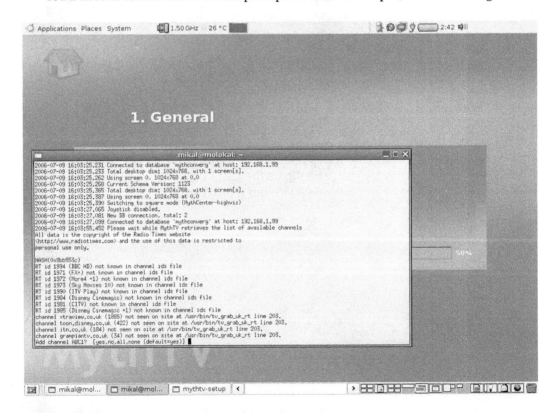

Figure 4-13. *The XMLTV grabber is configured in the terminal.*

The grabber will prompt for each of the channels it's aware of in your area one at a time, and you need to confirm which channels you receive and are interested in recording content from. For example, if you receive the Basket Weaving Channel but you're sure you'll never be interested in anything on it, just tell the grabber that you don't receive it.

Once you've finished with the text portion of the configuration, the graphical program will return to the video source menu as well.

Forcing a Guide Update

Now that you have the guide data set up, you need to populate the MySQL database that MythTV uses to store this information. You can force a guide data update using the mythfilldatabase command. This works at other times as well, such as if you suspect that your current guide data is inaccurate or not complete. Here's an example that uses a Zap2it subscription:

```
$ mythfilldatabase
2006-05-29 16:25:37.498 Using runtime prefix = /usr/local
2006-05-29 16:25:37.511 New DB connection, total: 1
2006-05-29 16:25:37.542 Connected to database 'mythconverg' at host: 192.168.1.99
2006-05-29 16:25:37.551 New DB connection, total: 2
2006-05-29 16:25:37.565 Connected to database 'mythconverg' at host: 192.168.1.99
2006-05-29 16:25:37.566 mythfilldatabase: Listings Download Started
2006-05-29 16:25:37.567 Updating source #1 (Comcast Mountain View) with grabber ➥
datadirect
```

You can see here that we have one lineup, which is the Comcast subscription in Mountain View, California. mythfilldatabase now checks whether it has guide data for this lineup and downloads what it doesn't have:

```
2006-05-29 16:25:37.568 Checking day @ offset 0, date: Mon May 29 2006
2006-05-29 16:25:37.625 Data is already present for Mon May 29 2006, skipping
2006-05-29 16:25:37.625 Checking day @ offset 1, date: Tue May 30 2006
2006-05-29 16:25:37.625 Refreshing data for Tue May 30 2006
2006-05-29 16:25:37.626 New DB DataDirect connection
2006-05-29 16:25:37.641 Connected to database 'mythconverg' at host: 192.168.1.99
2006-05-29 16:25:37.642 Retrieving datadirect data.
2006-05-29 16:25:37.642 Grabbing data for Mon May 29 2006 offset 1
2006-05-29 16:25:37.642 From Tue May 30 07:00:00 2006 to Wed May 31 07:00:00 ➥
2006 (UTC)
...
2006-05-29 16:25:47.490 DataDirect: Your subscription expires on 06/14/2006 ➥
10:57:06 PM
2006-05-29 16:26:00.422 Grab complete.  Actual data from Tue May 30 07:00:00 ➥
2006 to Wed May 31 07:00:00 2006 (UTC)
```

Here mythfilldatabase determines that it needs guide data for only one day (Tuesday, May 30, in this example):

```
2006-05-29 16:26:00.423 Main temp tables populated.
2006-05-29 16:26:00.423 Updating myth channels.
2006-05-29 16:26:00.464 Updating icons for sourceid: 1
```

```
2006-05-29 16:26:00.465 New DB connection, total: 3
2006-05-29 16:26:00.507 Connected to database 'mythconverg' at host: 192.168.1.99
2006-05-29 16:26:00.508 Channels updated.
2006-05-29 16:26:00.676 Clearing data for source.
2006-05-29 16:26:00.676 Clearing from Tue May 30 00:00:00 2006 to Wed May 31 ➥
00:00:00 2006 (localtime)
2006-05-29 16:26:00.677 New DB connection, total: 4
2006-05-29 16:26:00.694 Connected to database 'mythconverg' at host: 192.168.1.99
2006-05-29 16:26:20.102 Data for source cleared.
2006-05-29 16:26:20.102 Updating programs.
2006-05-29 16:26:24.313 Program table update complete.
2006-05-29 16:26:24.313 Checking day @ offset 2, date: Wed May 31 2006
2006-05-29 16:26:24.380 Data is already present for Wed May 31 2006, skipping
...
2006-05-29 16:26:26.306 New DB connection, total: 5
2006-05-29 16:26:26.454 Connected to database 'mythconverg' at host: 192.168.1.99
2006-05-29 16:26:26.906 Data fetching complete.
2006-05-29 16:26:26.907 Adjusting program database end times.
2006-05-29 16:26:27.884     0 replacements made
2006-05-29 16:26:27.888 mythfilldatabase: Listings Download Finished
2006-05-29 16:26:27.888 Marking generic episodes.
2006-05-29 16:26:28.286     Found 576
2006-05-29 16:26:28.286 Marking repeats.
2006-05-29 16:26:28.661     Found 1408
2006-05-29 16:26:28.661 Unmarking new episode rebroadcast repeats.
2006-05-29 16:26:29.012     Found 11
--16:26:29--  http://datadirect.webservices.zap2it.com/tvlistings/xtvdService
           => `/tmp/mythpost3AfLJT'
Resolving datadirect.webservices.zap2it.com... 206.18.98.160
Connecting to datadirect.webservices.zap2it.com|206.18.98.160|:80... connected.
HTTP request sent, awaiting response... 401 Unauthorized
Connecting to datadirect.webservices.zap2it.com|206.18.98.160|:80... connected.
HTTP request sent, awaiting response... 200 OK
Length: unspecified [text/xml]

  OK                                              34.39 MB/s

16:26:30 (34.39 MB/s) - `/tmp/mythpost3AfLJT' saved [613]

2006-05-29 16:26:30.018 DataDirect: NextSuggestedTime is: 2006-05-30T17:41:43
2006-05-29 16:26:30.025
=================================================================
| Attempting to contact the master backend for rescheduling.  |
| If the master is not running, rescheduling will happen when  |
| the master backend is restarted.                            |
=================================================================
```

```
2006-05-29 16:26:30.029 Connecting to backend server: 192.168.1.99:6543 (try 1 of 5)
2006-05-29 16:26:30.035 Using protocol version 26
2006-05-29 16:26:30.087 mythfilldatabase run complete.
```

At this point your MythTV system will have updated its guide data. Generally, the grabber will retrieve one new day at the far end of your 14-day program data window and also tomorrow. This is because by the time tomorrow becomes tomorrow, its data will be 13 days old and might have changed. If you think this implies that the conflict-avoidance algorithms might occasionally delay recording a program until a repeat that it later cancelled, you're right, and in Chapter 5 we'll show you how to manually mark programs that are critical to you to always record as early as possible.

Mapping the Video Source to a Tuner

Once you've added a video source to MythTV and run `mythfilldatabase`, the next step is to map your tuner to the video source. You do this so you can have different video sources on different tuners. Go to the Input Connections menu, and map the inputs on your capture card to your video sources. Figure 4-14 shows an example of our Hauppauge PVR card.

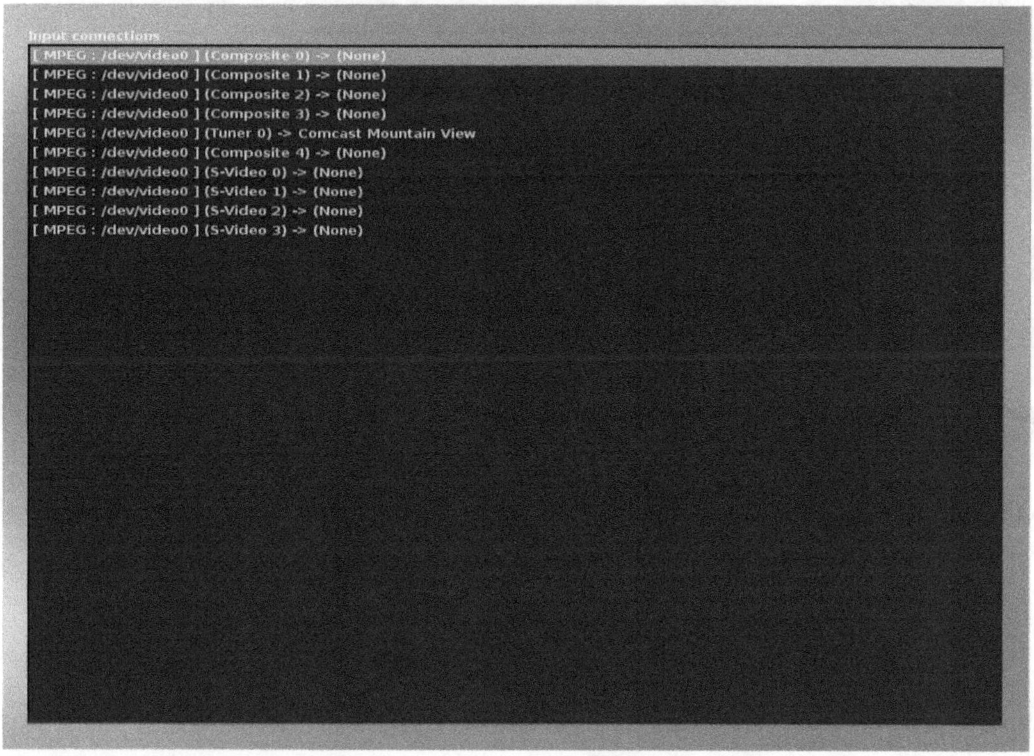

Figure 4-14. *The input connections configuration for a Hauppauge PVR card*

You can see from this example that we have only the tuner plugged into a video source, and it is connected to our Comcast cable TV. Which input to use on your card will vary based on what video source you're using, what type of card you have, and how you have the two connected. Your best bet is simply to make an educated guess and then use MythTV's Live TV feature to test that you have it right (though on the Hauppauge cards, it's most commonly Tuner 0, S-Video 0, and Composite 4). Figure 4-15 shows the configuration screen for our card.

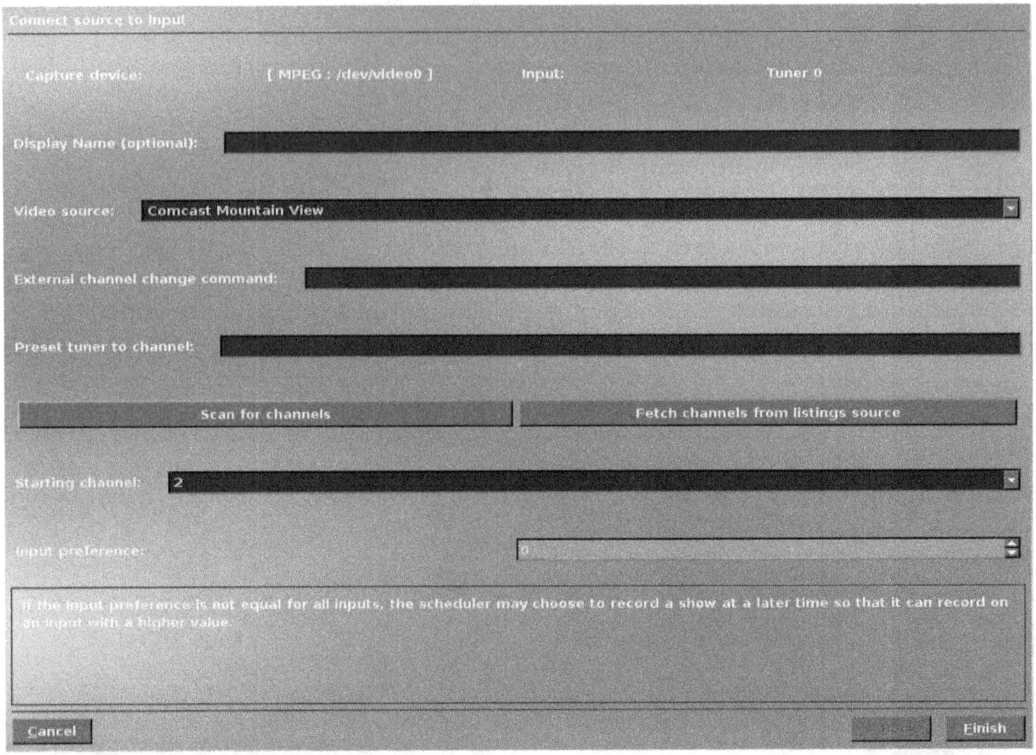

Figure 4-15. *Configuration for a Hauppauge PVR card*

You can see here that really you just need to tell MythTV which video source is connected to that input. We show how to configure the channel listing for the video source in the next section.

Getting Channel Listings

Finally, you need to make sure that the list of channels available to MythTV is correct. Select the Channel Editor menu item in the main menu of mythtv-setup. You'll be presented with the screen shown in Figure 4-16.

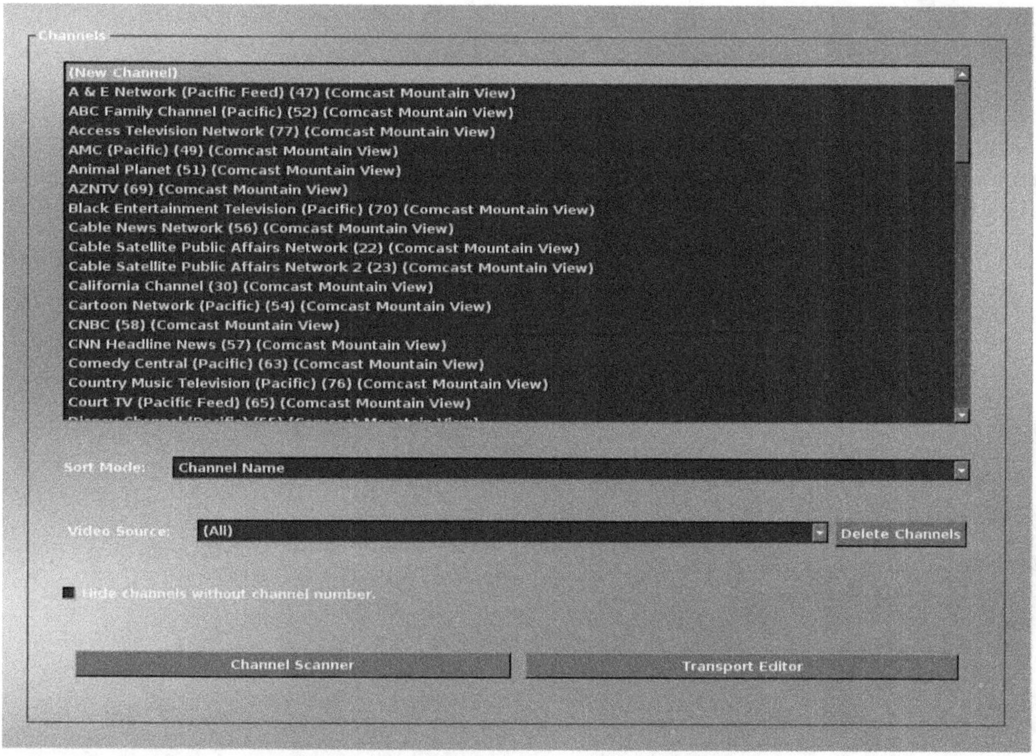

Figure 4-16. *The channel editor*

You should see a list of configured channels here already, because these were downloaded from the guide data source you configured earlier. If you don't see any channels listed here, then you have a couple of options. You can manually enter the channels, or you can have MythTV scan for them. The problem with the scanning technique is that the channel names will be missing. If there is a list of channels with names, you can check these against the supplied lineup for your program source (cable, and so on) or just flip through them and look.

Scheduling Recordings

Now that you have guide data, it's possible to start scheduling recordings. You can find shows to record in a variety of ways, each of them tailored to a different kind of usage. You can browse the TV guide using the program guide, you can search the guide data in a variety of ways, you can press Record while watching a show on Live TV, and finally you can schedule recordings in your web browser using the MythWeb plug-in. Each of these program discovery options helps you find shows. Once you have decided to record a show, you are presented with the same user interface regardless of how you came upon the show. We'll explore that user interface in the "Setting Up Your Scheduling Options" section.

All the recording options with the exception of Record from Live TV exist in the same place in the MythTV user interface. To navigate to them from the main menu, select Manage Recordings, as shown in Figure 4-17.

Figure 4-17. *The MythTV main menu*

And then select Schedule Recordings, as shown in Figure 4-18.

Figure 4-18. *The MythTV Manage Recordings menu*

You can see in the Schedule Recordings menu the various scheduling options we'll explore in this section, as shown in Figure 4-19.

Figure 4-19. *The MythTV Schedule Recordings menu*

The following sections cover each of these options one at a time.

The Program Guide

The program guide lets you browse through the shows that are currently on and select what you want to record or watch. Figure 4-20 shows a sample of what our online TV guide lists for today.

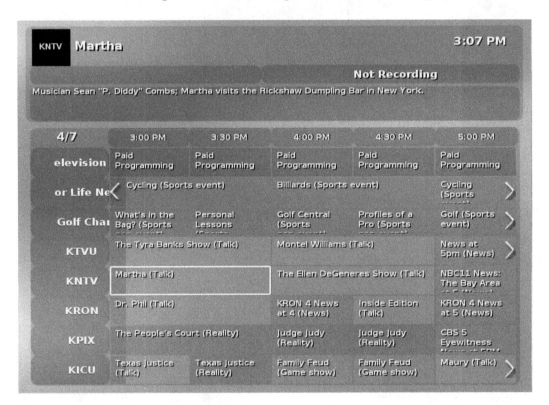

Figure 4-20. *The MythTV online TV guide*

You move around the guide as you would expect to with the arrow keys. When you have found a show that you want to record, select it by pressing OK on your remote or Enter on your keyboard. You'll then see the Recording Options screen, which we discuss in the "Setting Up Your Scheduling Options" section.

The Program Finder

The program finder is a more efficient way of finding shows if you know the name of the show. For example, let's look up the Discovery Channel's *Dirty Jobs*. The first step is to look up the first letter of the show's name, not including any leading *the*, as shown in Figure 4-21.

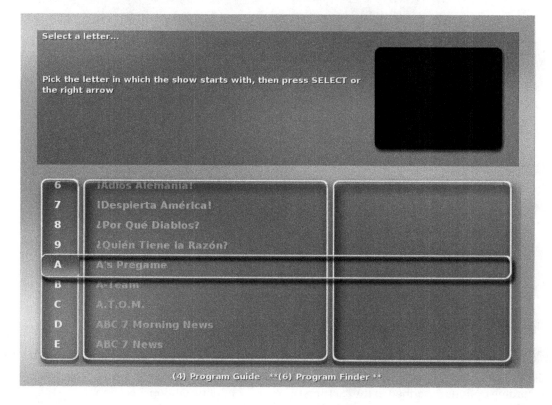

Figure 4-21. *Find the first letter of the show you want to record.*

Next scroll through the list of shows starting with that letter until you find the one you want. This is where this technique starts not to scale well. If a lot of shows start with the given letter, then you can be left hitting the down arrow for quite a while. Figure 4-22 shows *Dirty Jobs* selected.

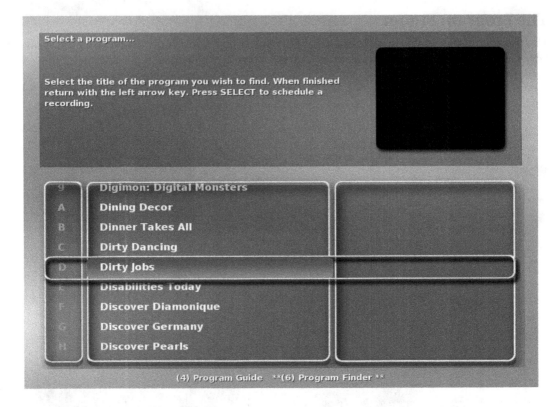

Figure 4-22. *Find the show you want to record.*

Finally, you'll see the times at which the show airs. You can choose any episode from the list and either record specific episodes or record all of them by selecting one and using the recording options described in the "Setting Up Your Scheduling Options" section.

Search Words and Lists

The next two search options present more powerful user interfaces but in return are harder to drive from a remote control. The first of these menu items, the Search Words option, shows you the menu in Figure 4-23.

We'll focus on the advanced option here, because the others are really just simplified variations on that advanced option. When you select one of these search options, you'll see a simple search interface, which shows you previous searches and lets you enter new searches, as shown in Figure 4-24.

Figure 4-23. *Ways you can search for words*

Figure 4-24. *The search dialog box*

You can either reexecute a previous search or enter a new one. Let's enter a new, advanced search, as shown in Figure 4-25.

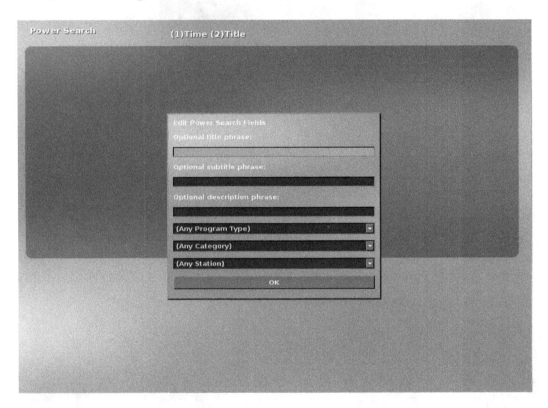

Figure 4-25. *Entering a new, advanced search*

You can see here that the advanced search lets you search by title, subtitle, other descriptive string, program type (movie, documentary, and so forth), genre (action, drama, and so forth), and channel. The other search options in the Search Words menu let you search for words in the title or keywords and for the names of the actors in the shows. If you are using a keyboard, you can just type in these text fields and use the arrows keys to navigate between them. If you are using a remote control, then press Select on a field to have an on-screen keyboard appear, as shown in Figure 4-26.

Once you've entered your search details and selected OK, the search results will appear. Figure 4-27 shows the search results that are automatically displayed if you select a previous search from the initial dialog box.

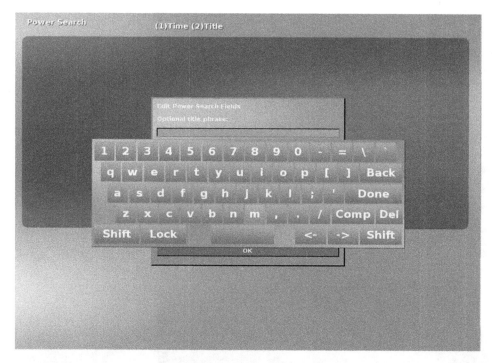

Figure 4-26. *The on-screen keyboard*

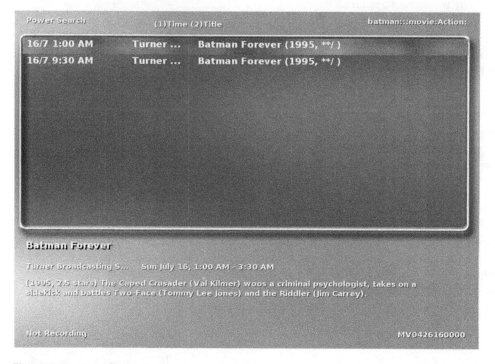

Figure 4-27. *Search results*

Selecting any of these results will once again take you to the recording options discussed in the "Setting Up Your Scheduling Options" section.

The other option here is searching by lists. Lists are commonly searched-on items, presented in alphabetical order. Figure 4-28 shows the list selection screen.

Figure 4-28. *The various list options*

These are all self-explanatory and provide useful ways of finding shows to record. Of all these listings, the one we use by far the most is the New Titles search—it's great for finding shows we haven't seen before.

Custom Record

The Custom Recordings option lets you set up searches that result in shows being recorded automatically based on them matching your search criteria, much like "wish lists" or "season passes" on TiVo. They're covered in the Chapter 5.

Manual Record

You can set up manual recordings for a fixed time on a given channel. The user interface is simple—define the channel, time, duration, and an optional title for the recording, and you're done. You might need to do this if, for some reason, you ever want to record two copies of the same program in the same time slot—MythTV won't let you do that otherwise. Another example of when you might need this is if there is a gap in your guide data but you still want to make a recording. Figure 4-29 shows a sample of the user interface.

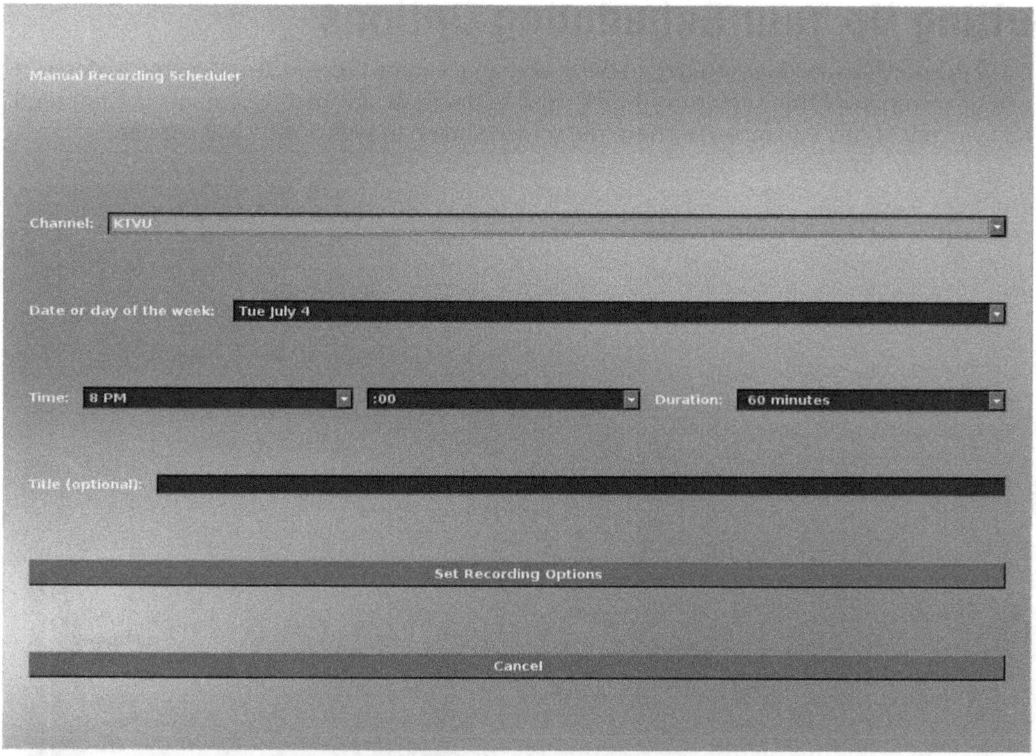

Manual Recording Scheduler

Channel: KTVU

Date or day of the week: Tue July 4

Time: 8 PM   :00   Duration: 60 minutes

Title (optional):

Set Recording Options

Cancel

Figure 4-29. *Setting up a manual recording*

Selecting Record on Live TV

When you watch Live TV using your MythTV frontend, MythTV records the show as if it were a scheduled recording, in the sense that the portion of the show you watched is placed in the same location as all your other recordings. Live TV recordings are expired automatically (see the Utilities / Setup ➤ Setup ➤ TV Settings ➤ General ➤ Live TV Max Age option for details), unless you tell MythTV to keep them. This form of recording doesn't give you any recording options, but you can use the one recording you make to set up a scheduled recording if needed.

Other Methods

One other method is available for scheduling recordings for MythTV. This is the MythWeb web interface. We won't discuss this in further detail here, because we cover MythWeb in Chapter 11 of this book.

Setting Up Your Scheduling Options

Most of the previous sections referred to the Scheduling Options screen. It's now time to describe the basic elements of that screen, and we'll explore the more advanced elements in Chapter 5. When you tell MythTV to record a show, the screen shown in Figure 4-30 will appear.

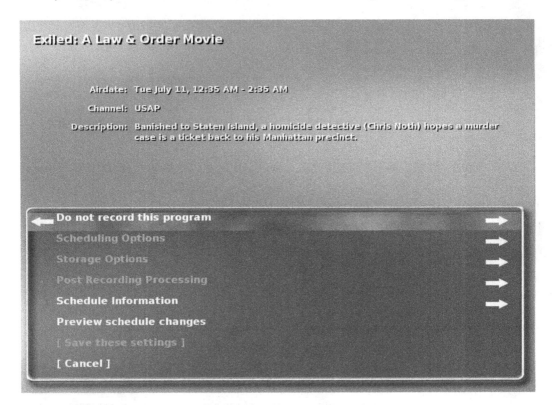

Figure 4-30. *The initial recording dialog box*

Most of the options are disabled here, but if you tell MythTV to record the show, the other options will be enabled. Figure 4-31 shows what the screen then looks like.

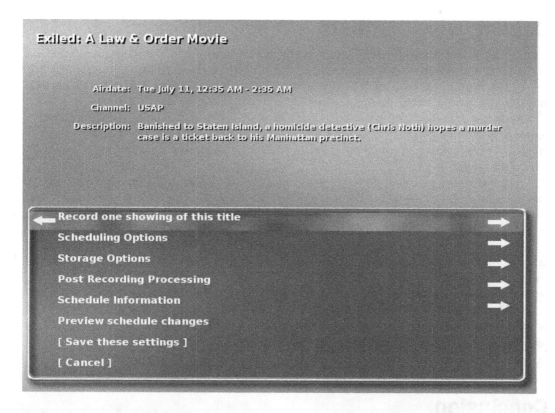

Figure 4-31. *Once you've selected Record*

We'll discuss all the options in this dialog box in Chapter 5, but the first option is the most important to getting you up and running. Here are the options in that menu:

- Do Not Record This Program

- Record Only This Showing

- Record One Showing of This Title

- Record in This Timeslot Every Week

- Record One Showing of This Title Every Week

- Record in This Timeslot Every Day

- Record One Showing of This Title Every Day

- Record at Any Time on This Channel

- Record at Any Time on Any Channel

These all seem fairly obvious. You can specify this specific airing or just one airing at any time. You can also record a variety of possible regular occurrences, depending on your preference.

Dealing with Conflicts

It's quite likely that you'll end up having scheduled recordings for two different shows that air at the same time (or, more generally, more programs than you have available tuners). This isn't as bad as it sounds, because MythTV can automatically resolve these conflicts in a variety of ways, or you can force your desired outcome.

As discussed in the "Setting Up Your Scheduling Options" section, you can configure MythTV to not record duplicates of previously aired episodes, and this is something we recommend enabling. This means you can avoid missing out on the new show by recording a show you have already seen.

If two recordings are competing for the same time slot, then it is also possible that MythTV can find one of the shows being aired at a different time and record that show at this other time instead, thus freeing the tuner to record the other show in the original time slot. To enable this to happen, you're best off setting shows to the Record at Any Time on Any Channel option.

Next, you can enable MythTV to record more than one show at once by adding more tuners. You can find more details on how to configure this and the implications that having more tuners running will have for your hardware requirements in Chapter 10.

Finally, if the recording options haven't been able to solve your tuner shortage and you can't add more tuners for some reason—for example, you have budgetary constraints or a lack of expansion slots—then you can also set priorities for different shows. This allows you to at least specify which show is more important than the other. You do this with the recording priority setting, which was described in Chapter 5.

Conclusion

Thanks for bearing with us through this chapter. Getting guide data working and learning how to record shows are not the most exciting topics in this book, but they are both vital to having a working MythTV system. We've now laid the groundwork for some of the more exciting projects later in this book.

The next two chapters cover the issues surrounding advanced TV recording, such as how to automatically remove commercials, how to transcode to other video formats such as those accepted by Apple iPods and Sony PlayStation Portables, before moving on to discuss other features of MythTV available in the user interface.

CHAPTER 5

■ ■ ■

Performing Advanced TV Recording

In Chapter 4 we discussed the basics of recording television with MythTV. Now that we have those basics covered, we're ready to cover the more advanced aspects of recording television. These advanced aspects include the recording options we didn't discuss in Chapter 4: setting up automated commercial detection and skipping, autoexpiring recordings to save on disk space, *transcoding* (converting video to other formats), automating transcoding, and adding user-defined jobs to MythTV.

Exploring the Advanced Recording Options

We referred to a few recording options in Chapter 4 that we never explained in full. The time has come to discuss those options. First we'll discuss custom recordings, and then we'll discuss the various scheduling options that are available.

Custom Recordings

Custom recordings give you the ability to express complicated requirements for which shows you want recorded. Additionally, they provide the "wish list" functionality that is offered by commercial PVR solutions such as TiVo. A simple example is if you want to record all programs that contain a given word in their titles. You can find custom recordings using the Manage Recordings ➤ Schedule Recordings ➤ Custom Record menu in MythTV, which takes you to a screen that looks like Figure 5-1.

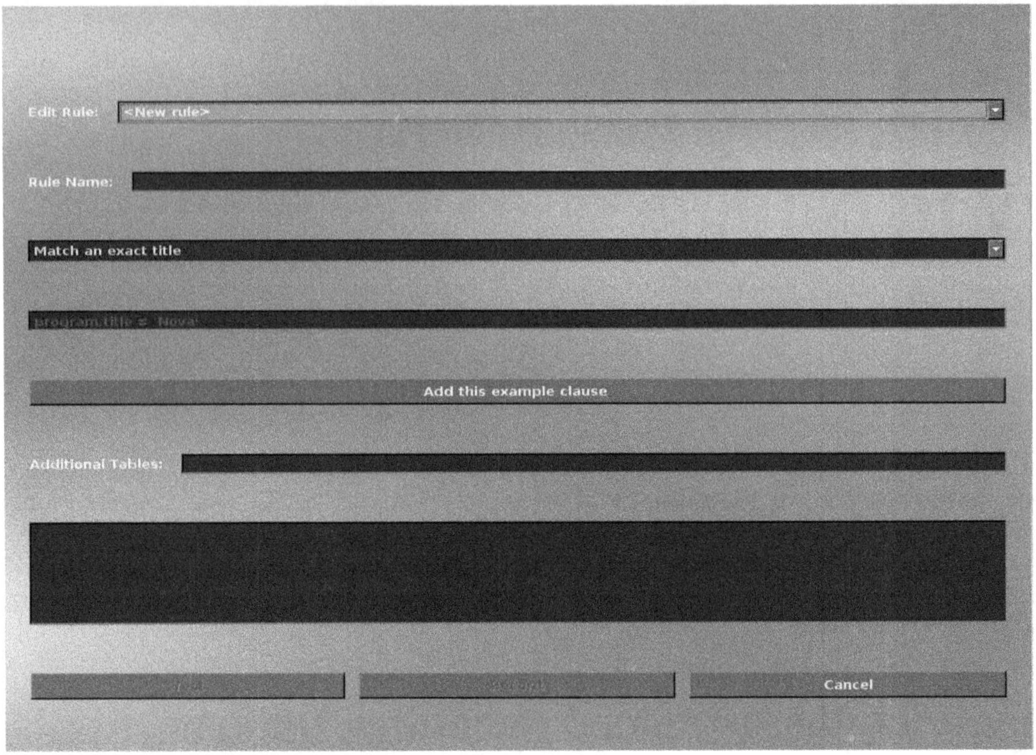

Figure 5-1. *The custom record screen*

This screen lets you edit *rules*, which are essentially saved searches that MythTV executes each time it downloads more guide data. You can also create new rules here as well. Once you've either chosen a new rule or selected an existing rule, you specify the title for this rule. Rules are built up from a series of one or more clauses, and these clauses are boolean ANDed together to form the complete rule. These rules are expressed in SQL, which might not be familiar to you if you haven't done a lot of database programming.

WHAT IS SQL?

Modern relational databases use Structured Query Language (SQL) to express the operations that you can perform on the data in the database. The operations usually supported by these relational databases are as follows:

- Storing new data (INSERT)

- Finding data (SELECT)

- Removing data (DELETE)

- Merging data from more than one table (JOIN)

SQL is more complicated than that, but all you need to know for the purposes of this discussion is that the clauses you write for these rules are used in SELECT statements, and anything you can do in a WHERE clause for a SELECT should work. Of course, some modern databases aren't relational, but those don't use SQL.

A variety of sample clauses are available in the next section of the screen. Examples include the following:

- Match an Exact Title

- Match an Exact Episode

- Match Words in the Title

- Match in any Descriptive Field

- Limit by Category

- All Matches for a Genre (depends on guide data source)

- Limit by Rating (depends on guide data source)

- New Episodes Only

- Exclude Unidentified Episodes

- Category Type (movie, sports, TV show, and so on)

- Limit Movies by Year of Release

- Minimum Star Rating for Movies

- Person Named in the Credits

- Only on a Specific Station

- Exclude One Station

- Match Related Callsigns

- Only on Channels Marked As Favorites

- Only on Channels from a Specific Video Source

- Only on Channels Marked As Commercial Free

- Only Shows Marked As HDTV

- Anytime on a Specific Day of the Week

- Only on Weekdays

- Only on Weekends

- Only in Primetime

- Not in Primetime

- Multiple Sports Teams

- Sci-Fi B Movies

- SportsCenter Overnight

- Movie of the Week

- First Episodes (depends on the data source)

You can see that this is a pretty good selection of example clauses on which you can base your custom recordings. Let's look at one of the more complete examples, which will give you more a feel of the type of SQL that you can write for these clauses (see Figure 5-2).

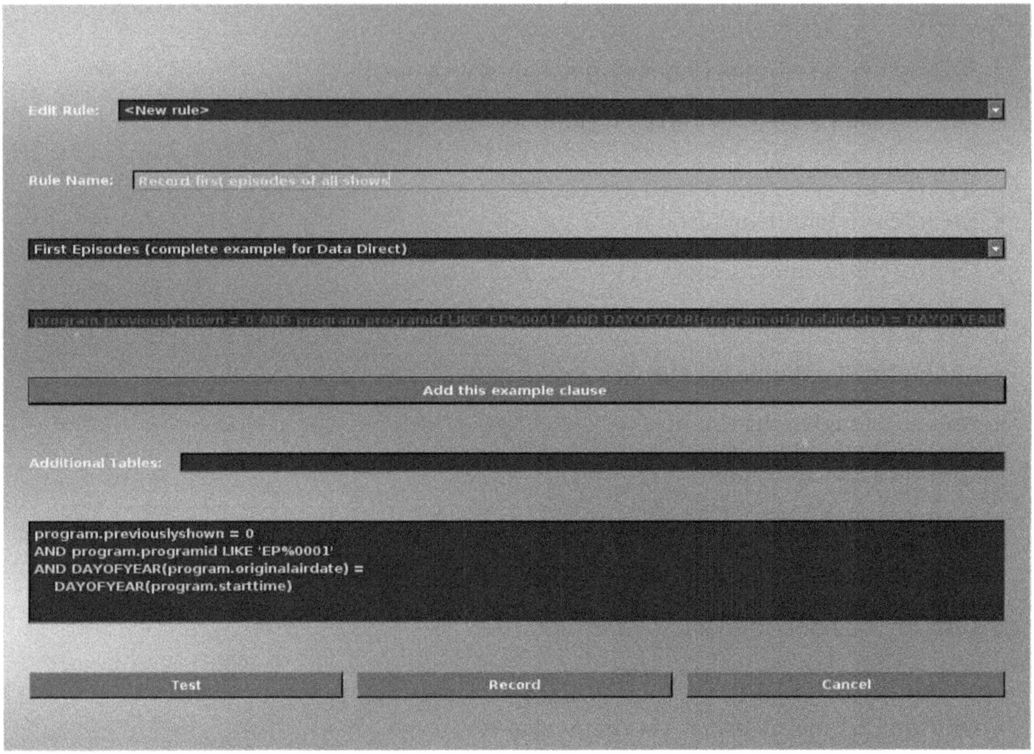

Figure 5-2. *An example clause*

The SQL clause here is as follows:

```
program.previouslyshown = 0
AND program.programid LIKE 'EP%0001'
AND DAYOFYEAR(program.originalairdate) = DAYOFYEAR(program.starttime)
```

This clause will record all the first, or *pilot*, episodes of all new shows aired. The previous examples should provide a good basis for you writing your own queries.

Recording Profiles and Transcoding Options

Recording profiles are how you set up your preferences for the quality of video to be stored for various purposes on your MythTV machine. You configure recording profiles using the

Utilities / Setup ➤ Setup ➤ TV Settings ➤ Recording Profiles menu. We have two profiles listed: the one for the hardware MPEG-2 encoder in our TV capture card and the built-in transcoders MythTV supports. It's possible you might have more depending on what capture card you are using. You can also create your own recording profiles if you want to do so.

Generally, you create different recording profiles if you want different qualities of recording. For example, you might not care about having music videos in a high-quality video format, because you don't watch the videos as much as you listen to the music. On the other hand, you might particularly care that documentaries are in the highest possible video quality because you're particularly interested in the scenery. Remember, the higher the video quality, the more space it takes on disk. Additionally, if you have a video capture card that supports MPEG-2 compression in hardware, the profile "programs" the encoder on the card for recordings; the end effect is the same, though the mechanism is different. Doing this compression in hardware is advantageous because it means less data is transferred over the PCI or USB bus per frame, which means more tuners can be supported on that bus. It also frees up the CPU to perform other tasks.

Under the hardware MPEG-2 recording profile, we have the following options: Default, Live TV, High Quality, and Low Quality. These are different uses of the parent recording profile based on various needs. Figures 6-3 through 6-6 show a complete example of the high-quality recording profile as we have it configured.

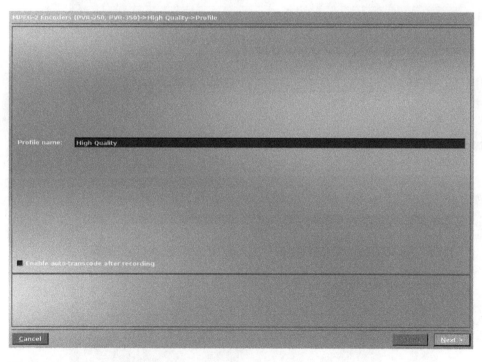

Figure 5-3. *Recording profile setup dialog box screen 1*

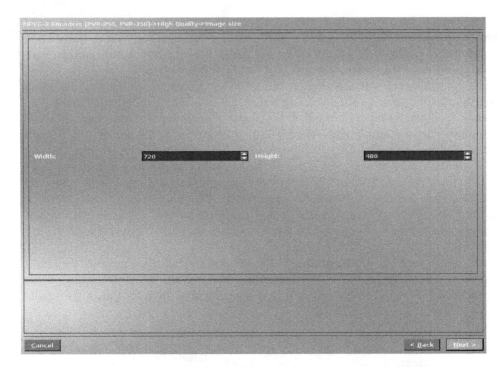

Figure 5-4. *Recording profile setup dialog box screen 2*

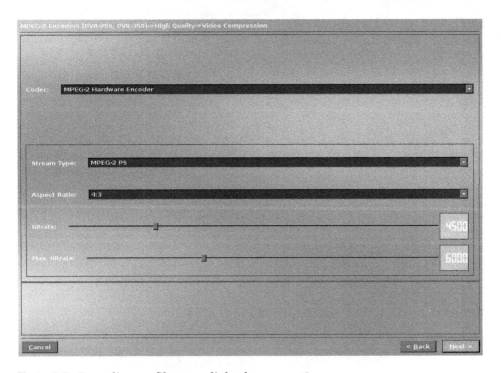

Figure 5-5. *Recording profile setup dialog box screen 3*

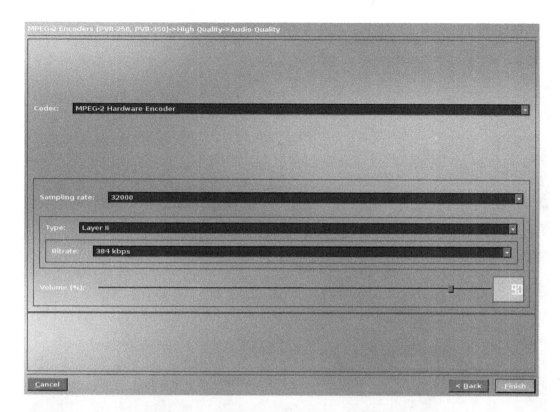

Figure 5-6. *Recording profile setup dialog box screen 4*

You can see here the functionality a recording profile gives you and that the setup wizard is pretty wasteful of screen resources. Options include the resolution to record at, the bit rate for audio, and the compression codecs to use for video and audio. To reduce the amount of disk space needed for a recording, you can use a lower resolution and lower bit rate, although doing this will reduce the overall quality of the recording.

Transcoding profiles are pretty much the same as the other group of recording profiles but are applied when a video is being transcoded after being recorded, instead of at recording time itself. You will learn more about transcoding when we discuss `mythtranscode` later in this chapter.

When you chose to record or transcode a video, you will be asked to select a recording profile, and the recording profile you select will have its settings applied to the job. You can therefore think of recording profiles much like saved preferences for these tasks.

Playback Groups

You can also set simple preferences for how MythTV behaves during the playback of recordings using playback groups. Options configurable for a playback group include the distance that a skip moves you in the recording (when you hit the right or left arrow) and at what speed the recording is played. For example, perhaps you're interested only in scanning sports broadcasts and want them to play back at twice the normal speed. You can set up a playback group for sports programs, and then this will automatically happen.

You configure these groups using the Utilities / Setup ➤ Setup ➤ TV Settings ➤ Playback Groups menu. You'll see a list of existing playback groups and the option of creating a new playback group. When creating a new playback group, you get the options shown in Figure 5-7.

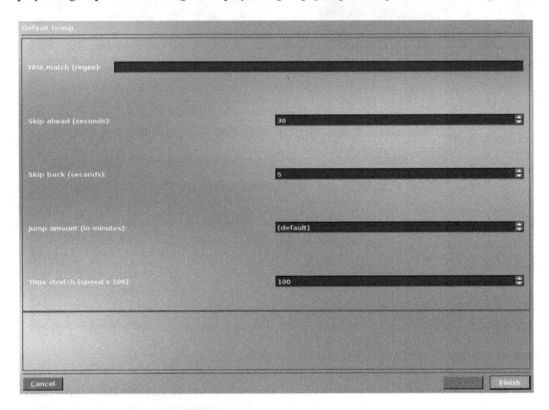

Figure 5-7. *Creating a new playback group*

You can set this playback group to match titles based on regular expressions and then you specify how far to skip forward and backward for the skip buttons, as well as how far to jump for the jump button and how fast to play the video. Table 5-1 lists some examples of regular expressions.

Table 5-1. *Some Example Regular Expressions*

Regular Expression	Matches
NFL	All shows with *NFL* in their title
(CNN|News)	All shows with either *CNN* or *News* in their title.

Recording Options

A variety of other options are available when you select to record a program. You might recall the record screen from Chapter 4 (see Figure 5-8).

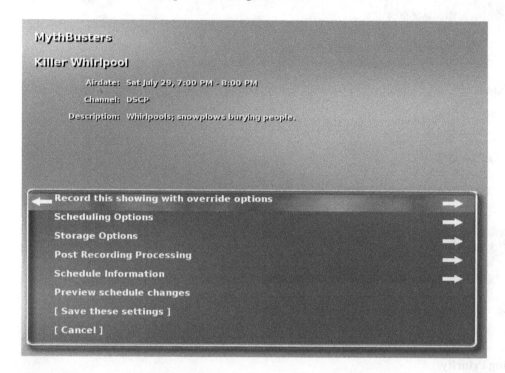

Figure 5-8. *Once you've selected record*

We've previously discussed the simplest recording option—when to record the program. Examples included once a day, at a specific time, or at any time on any channel. This screen has other options, though, and we haven't covered those yet. Let's work through them now.

Scheduling Options

The scheduling options affect the decision of whether to record a given showing of a program, some in ways that aren't entirely obvious. Figure 5-9 shows the options that are available.

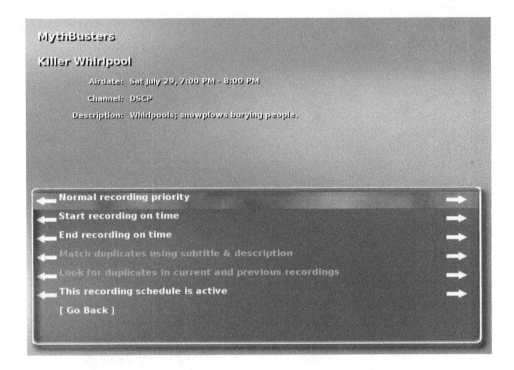

MythBusters

Killer Whirlpool

Airdate: Sat July 29, 7:00 PM - 8:00 PM

Channel: DSCP

Description: Whirlpools; snowplows burying people.

◀ Normal recording priority ➡

◀ Start recording on time ➡

◀ End recording on time ➡

◀ Match duplicates using subtitle & description ➡

◀ Look for duplicates in current and previous recordings ➡

◀ This recording schedule is active ➡

[Go Back]

Figure 5-9. *The main scheduling options menu*

Recording Priority

The recording priority is a number from –99 to 99. The default is zero, and the higher the number, the more likely that a show will be recorded. In fact, four priority numbers exist; the one we are setting here is for the program in general. You can also set priorities for the type of recording (for example, single recordings versus any time on any channel), for the channel that is being recorded, and for a given video input to the machine. You set this using Utilities / Setup ➤ Setup ➤ TV Settings ➤ Recording Priorities. The scheduler looks at all these priorities when deciding whether to record a given showing of a program.

We already discussed conflict resolution in Chapter 4, but if you find that the automated conflict resolution doesn't work the way you want or that you prefer a few particular shows more than many others, then the priority options can be quite useful. We use the recording priority to ensure that shows we're particularly interested in are always recorded before shows we don't care as much about. In other words, we have important shows, and we have shows that we will record if there is spare tuner capacity.

Recording Time Padding

The next two options allow you to specify padding for the start and end of the recording. This can be useful if you know that the network airing a show tends to start the show a little bit earlier than advertised or if the show tends to run past the scheduled ending time. An example is *The Simpsons* on Fox in the United States. It always seems to start a minute or so early, so adding padding stops us from missing the first little bit of the show.

Of course you'll encounter a trade-off here. By adding padding to a particular recording, you are stopping the MythTV machine from recording other shows on that input, because it won't be able to record other shows that conflict with the overrun. You should bear this in mind when you add padding.

Duplicate Matching

MythTV can use a couple of duplicate matching techniques to make sure it doesn't record a given episode of a show more than once. These options are as follows:

- Match Duplicates Using Subtitle and Description

- Match Duplicates Using Subtitle

- Match Duplicates Using Description

- Don't Match Duplicates (which records all airings)

Most shows will have a title, a subtitle, and a description. If you refer to Figure 5-9, then you can see that the title for the show in this example is *MythBusters*, the subtitle is *Killer Whirlpool*, and the description is "Whirlpools; snowplows burying people." In some cases, the guide data provided by your guide data source might not provide reliable information for some or all of these fields. MythTV assumes that the title is reliable and lets you tune which of these other fields MythTV trusts for the purposes of duplicate detection. If the title field is unreliable (for example, if the guide data uses slightly different names for the same show over time), try using the custom recording schedules to match keywords in the title instead of the entire title.

Finally, you can specify how MythTV decides that a show has been previously recorded, which affects whether the show is recorded again. The options are as follows:

- Record New Episodes Only. (This option looks at the original air date field in the guide data and won't record shows where that date is more than 14 days older than the air date of the show.)

- Look for Duplicates in Current and Previous Recordings. (A current recording is one that is still available for playback. A previous recording is one that has been recorded before but is no longer on disk; in other words, it was deleted for some reason.)

- Look for Duplicates in Current Recordings Only.

- Look for Duplicates in Previous Recordings Only.

Whether the Recording Schedule Is Active

You can deactivate a recording schedule if you want that show to not be recorded for a while. For example, if you're going to be away from home for a month and you don't want the daily news recorded for that time because you want to save the disk space for shows you care more about, then you can deactivate that recording schedule while you are away.

Tip A *recording schedule* is a set of rules that determines whether a show should be recorded. A *scheduled recording* is the result of a "yes" answer to that question during the scheduling process. The two terms are similar; don't be confused by them.

Storage Options

A variety of options affect how MythTV stores the programs that are recorded, and you can alter these options using the Storage Options menu. You can of course set different options for different programs, which helps you tailor MythTV's behavior to your needs. Figure 5-10 shows the configuration screen with all the options.

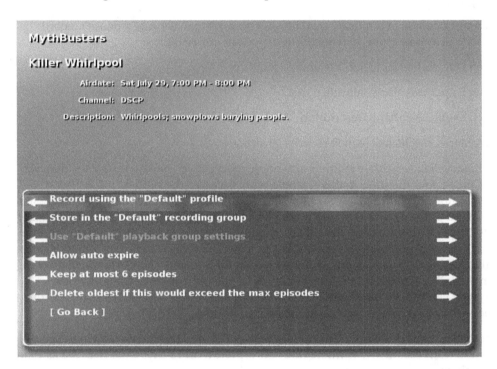

Figure 5-10. *The main Storage Options menu*

First, you can select the recording profile you want to use for this show. Refer to the earlier "Recording Profiles and Transcoding Options" section in this chapter for more information about how to set up a recording profile. Next, you can select the recording group to use. *Recording groups* are a way of organizing entries in the Watch Recordings menu, as well as the MythWeb web interface. You can select a recording group from the menu or create a new recording group right away by selecting the new option from the list of recording groups. A recording group is just a descriptive string used for display and selection, so you can enter pretty much anything. Figure 5-11 shows what the dialog box for creating a new recording group looks like.

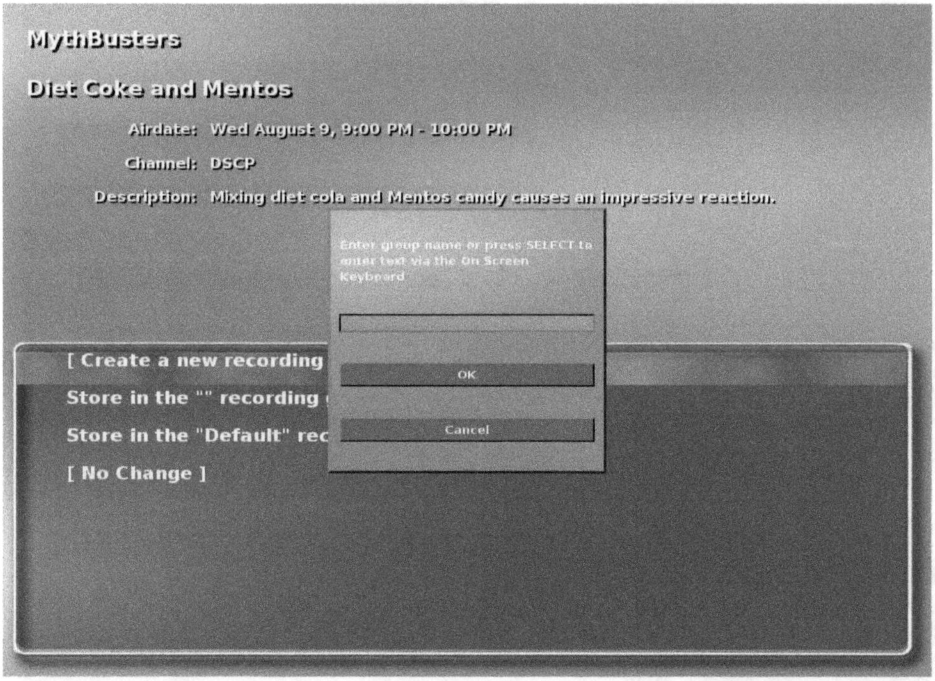

Figure 5-11. *Creating a new recording group*

You might notice that the background for that dialog box is a little different from the other examples. That is because if you hit Enter (OK on the remote) on any of these menu selectors, you get a dialog box with the full list of options, instead of having to use the left and right arrows to see the options.

Similarly, you can change the behavior of some aspects of playback with playback groups. We discussed these earlier in this chapter as well.

Autoexpiry is the process of deleting recordings automatically when the MythTV backend has reached the minimum amount of free disk space as configured in the general settings. Figure 5-12 shows a snapshot of the relevant settings page, which can be found at Utilities / Setup ➤ Setup ➤ TV Settings ➤ General ➤ Extra Disk Space (on the third page of the wizard).

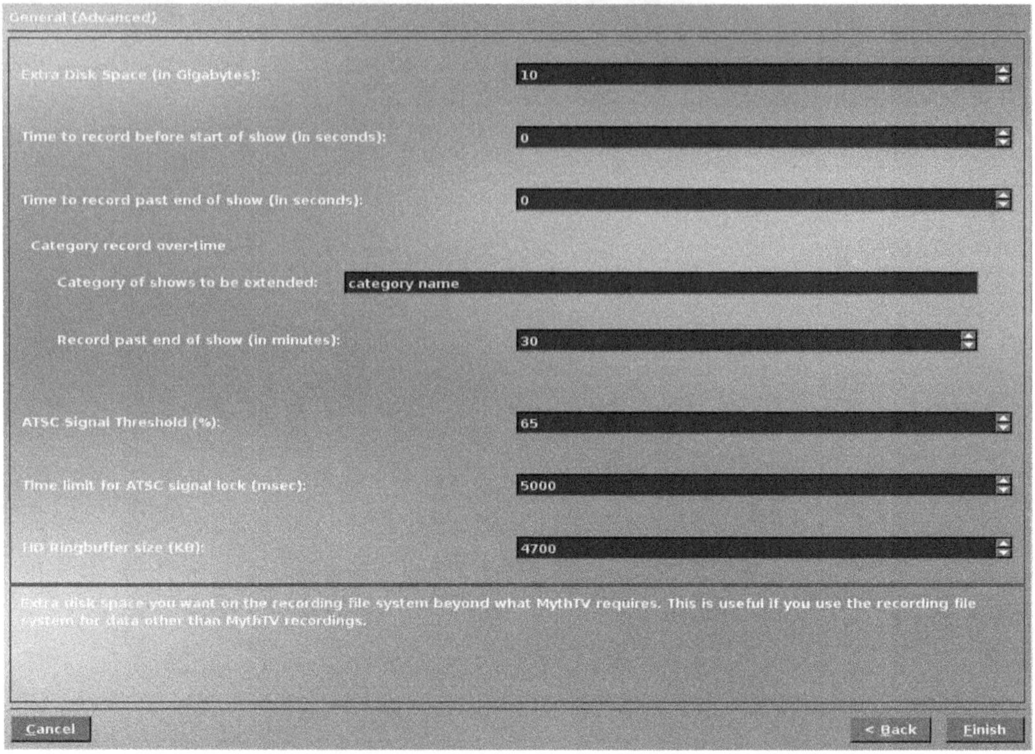

Figure 5-12. *Setting the minimum free disk space*

You can set autoexpiry to delete the oldest shows first or the shows with the lowest priority, although this latter option helps only if you have set different priorities for your recordings. You can toggle this behavior in the general TV settings wizard as well. The autoexpiry option here allows you to either enable or disable the autoexpiry setting of this specific show.

Returning to the recording options for just one show, you can specify how many episodes of the show to keep, from 1 to 100, or you can specify to keep an unlimited number. If you are going to exceed this limit, then you can select whether you want the oldest episode to be automatically deleted or whether you just don't want new episodes to be recorded.

Finally, if you archive lots of older recordings and want to keep them, you can turn off autoexpiration and watch your disk space manually. If you get low on disk space, then you can manually delete or archive to a DVD any programs you don't need to keep online. (We cover how to archive to DVD in Chapter 12.) Another option is to move the video files for some recordings to another disk and then create a symlink from the old location to the new location. Turning off autoexpiry can be a little more complicated in newer releases, because this affects how Live TV recordings are expired as well. Then again, if you're archiving a lot of shows, you might not be that interested in Live TV anyway.

Post Recording Processing

MythTV can perform a variety of operations on a video once it has finished recording. Figure 5-13 shows the dialog box where you can configure these operations.

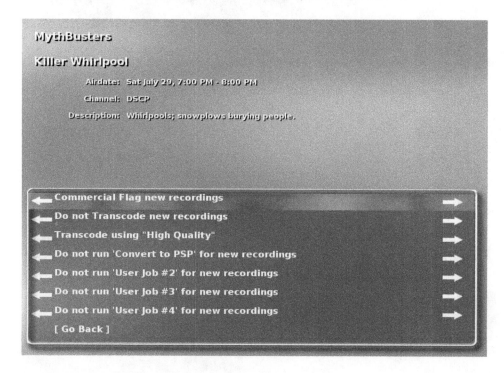

Figure 5-13. *Optional processing after recording has finished*

Commercial flagging is the process of detecting and marking commercials. This is then used for automatically skipping the commercials during playback, and the commercial markers can also be converted to cut points, which can be used during transcoding to reduce the length (and therefore size) of videos. You can configure the parameters used to detect commercials in Utilities / Setup ➤ Setup ➤ TV Settings ➤ General, on the second screen of the wizard. We discuss this more later in the "Detecting and Skipping Commercials" section. Figure 5-14 shows a sample of the options.

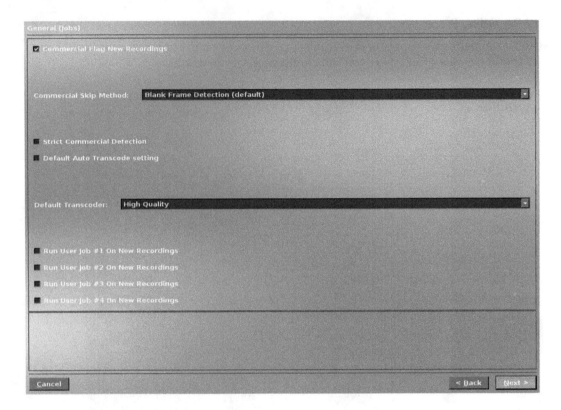

Figure 5-14. *Commercial detection options*

You configure whether post-recording transcoding occurs in this screen. Transcoding is often done in this context to save space on disk—both by converting the program to a more space-efficient compression scheme or smaller frame size and by recovering the extra space taken by removed commercials. You should note that transcoding to more aggressive compression schemes can also mean that the videos don't playback as well on frontends with limited CPU speed. For example, we have had trouble getting .mp4 transcoded videos, which are much smaller, playing well on the Xbox frontend that we describe in Chapter 8. You configure transcoding options in the recording profile settings, which were discussed earlier in this chapter in the "Recording Profiles and Transcoding Options" section. If you enable transcoding, then the next option specifies which transcoding settings to use.

Finally for this screen, you can run up to four *user jobs* for a given recording. These might be transcoding to other platforms such as Sony PlayStation Portable or Apple iPod video (which is the other meaning of the term *transcoding*) or can be other arbitrary tasks such as moving the video to a different location or emailing you. You'll note that Figure 5-13 doesn't have a user job 1. This is because you can replace the names in the user interface with your own descriptive text, which we have done for user job 1. We discuss how to set up user jobs to perform transcoding in much more detail in the "Transcoding" section.

Schedule Information

MythTV displays the schedule information about the current program at the top of the screen on the recording controls screen. Figure 5-15 shows the different tasks you can do here.

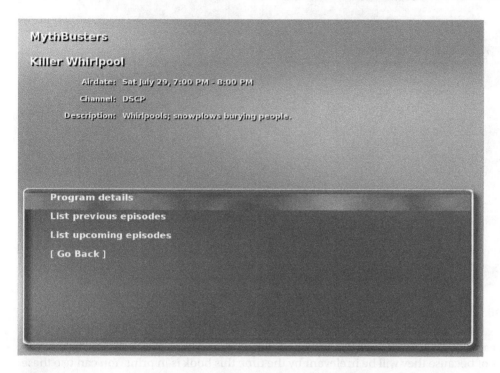

Figure 5-15. *The schedule options from the recording menu*

You can see program details here. This is just a view of information about the show you're recording on this screen. Figure 5-16 shows the example for the *MythBusters* episode we have been using in this chapter.

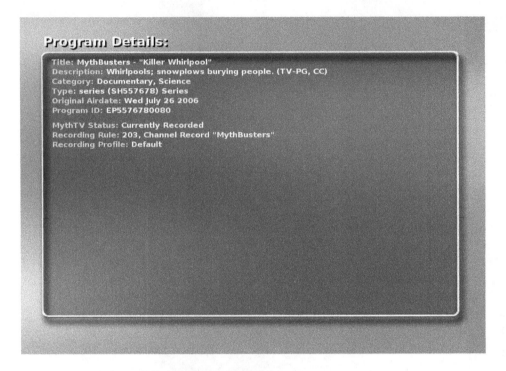

Figure 5-16. *Program details for a MythBusters episode*

You can also get a list of previous and upcoming episodes here, which we won't show you examples of because they will be irrelevant by the time this book is in print. You can use these lists to get an idea of how frequently the show airs and what sort of repetition options you should set for the recording.

Preview Schedule Changes

The preview option will show you how the upcoming recordings for your MythTV installation will change based on asking to record this show. This can be useful if another show that you are more interested in has been dropped off the upcoming recordings list because of this new recording. In that case, you might consider manually forcing the other show to be recorded or changing the priorities of the recordings so the right one is recorded.

Detecting and Skipping Commercials

As mentioned a few pages ago, *commercial flagging* is the process of detecting and marking commercials. This is then used for automatic skipping of the commercials during playback, and the commercial markers can also be converted to cut points, which can be used during transcoding to reduce the length (and therefore size) of videos. We'll discuss those editing options in a second, but now we'll talk about the commercial detection options and what they do.

You can configure the parameters used to detect commercials in Utilities / Setup ➤ Setup ➤ TV Settings ➤ General, on the second screen of the wizard. Figure 5-17 shows a sample of the options available.

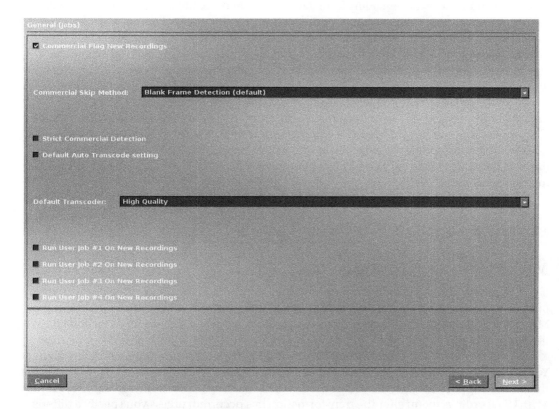

Figure 5-17. *Commercial detection options*

The most important options to know about here are the commercial detection methods and the strict option. The commercial detection options are as follows:

- Blank Frame Detection (the default)

- Blank Frame Detection + Scene Change

- Scene Change Detection

- Logo Detection (the station logo, that is)

- All of the Above

These options are all pretty self-explanatory. For them to work, you need to have the Commercial Flag New Recordings check box selected. The strict mode for commercial detection should usually be on, but you can experiment with turning it off if you have troubles with commercials not being detected.

Transcoding

We've already briefly mentioned transcoding in this chapter. *Transcoding* is the process of converting from one digital encoding format to another. In the context of this book, we use the term almost always to refer to changing the format of a video file—for example, changing the native NUV or MPEG-2 format that MythTV uses into the MPEG-4 format that a PlayStation Portable uses or the Apple QuickTime format that an Apple iPod video uses. More generally, you can use the term for other encoding changes, such as converting between image storage formats such as PNG and JPEG.

It is also possible that you will want to transcode from one format to the same format. This might be because you want to change the resolution of a video to save space or because you want to do something like remove commercials from a recording to make it smaller. Note, however, that transcoding is lossy—if you throw information away (like those commercials or the extra pixels that made the video a higher resolution), there is no way to get them back again— they are gone forever.

You have three ways to transcode the videos made by MythTV: MythTV's internal transcoding options; manual transcoding using nuvexport; and automatic transcoding of certain recordings, again using nuvexport. We'll cover all three in this chapter.

MythTV's Internal Transcoding

If you want to transcode a video to remove either commercials or other video that you have marked as not wanting anymore via a cut list and you want the video to end up back in the MythTV recordings menu, then you should use MythTV's built-in transcoding functionality. We discuss how to edit cut lists more in Chapter 6, so we'll gloss over how to make them for now.

MythTV's transcoding functionality is exposed via the mythtranscode command-line program. You can use this program to convert from whatever format a recording is already in to a format that is still supported by MythTV. This means the recording will continue to show up in the MythTV recordings menu after the transformation has occurred, unless you specify a different filename for the new file, at which point MythTV won't know about the new file.

We'll first show you how to run mythtranscode manually, and then we'll talk about how to make that more nicely integrated with the rest of the MythTV user interface. mythtranscode needs to know which file to transcode. You can do this in one of two ways: you can specify the filename of the recording to transcode; alternatively, you can specify the channel identifier and the start time of the program to transcode, and mythtranscode will find the file for you. We'll assume you know the name of the file to transcode, because at the time that we go to integrate this with the rest of the MythTV user interface, it is really easy to get the video filename. Here's an example of specifying the filename to transcode:

```
mythtranscode -i /data/mythtv/1073_20060902103000.mpg
```

You might want to tell mythtranscode which recording profile to use for the output file. This is how you specify the resolution, and so forth, to use for the new video file. You do this with the --profile command-line option, which takes a profile number. This is an advanced feature, so we'll leave looking up a profile number in the MySQL database as an exercise for the reader.

You'll probably want to specify that the cut list is honored when transcoding the video, because that's probably the reason you're using mythtranscode. You do this with the --honorcutlist command-line option, which doesn't take any arguments.

Finally, if you're transcoding from MPEG-2 to MPEG-2, then use the --mpeg2 command-line option to make the conversion as lossless as possible—you do this by reencoding only those frames where the cut list will result in the relevant keyframe being removed. If the keyframe remains, then the keyframe and the other frames are just copied. This will help ensure that the highest-quality video is produced in the end. You can tell whether you're storing video in MPEG-2 by looking at the recording filenames. If the filename ends in .mpg, then it's an MPEG-2 file. An common example is video recorded by the Hauppauge family of capture cards, which is converted to MPEG-2 by the capture card itself and therefore saved in MPEG-2 on disk as well.

Now let's see a sample mythtranscode command line. This example will reencode the named recording using the cut list to permanently remove the video marked as being unwanted by the cut list. It will also reencode from MPEG-2 to MPEG-2:

```
$ mythtranscode -i 1073_20060902103000.mpg --honorcutlist --mpeg2
2006-12-31 14:58:37.518 Using runtime prefix = /usr/local
2006-12-31 14:58:37.625 New DB connection, total: 1
2006-12-31 14:58:37.672 Enabled verbose msgs: important
2006-12-31 14:58:37.701 New DB connection, total: 2
Mux rate: 6.49 Mbit/s
```

One of the reasons that mythtranscode's user interface is a little obtuse (who knows the channel identifiers on their systems?) is because it's intended to be used with the "user jobs" feature of MythTV. In short, you need to craft a command line like we did previously and then use the variable names described in Table 5-2 in the "Setting Up the User Jobs" section later in this chapter to refer to the video to process.

Here's an example of the previous command line but suitable for a user job:

```
mythtranscode -i %FILE% --honorcutlist --mpeg2
```

You then put this command line into one of your user jobs via the user jobs configuration screen in the mythtv-setup application. Go to the General menu, and then hit Next until you end up at the user jobs page, as shown in Figure 5-18.

Tip You might find that you don't have a mythtv-setup program. That's probably because on some distributions it is named mythtvsetup.

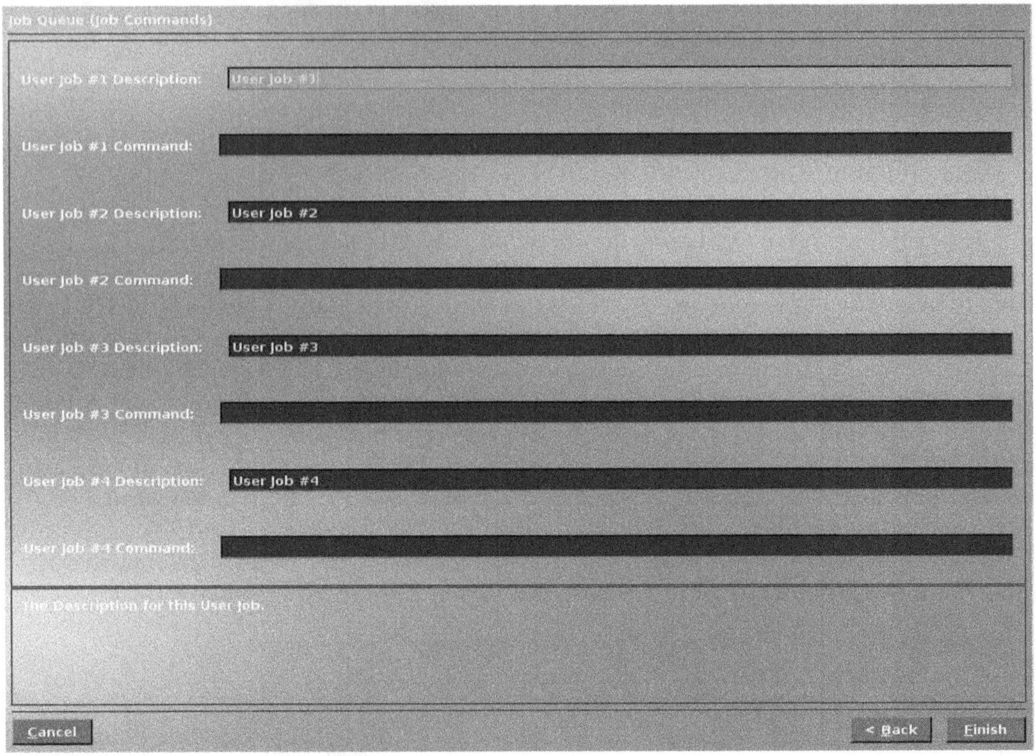

Figure 5-18. *The default user jobs configuration screen*

Here you need to give a meaningful name to the user job, for example "permanently remove cut list items," and then enter the previous command line in the command textbox. You can read more about setting up user jobs in the "Setting Up the User Jobs" section later in this chapter.

nuvexport

We do manual video transcoding with a command-line application called nuvexport. We're using version 0.4 from https://svn.forevermore.net/nuvexport/, which didn't work without some simple modifications. We'll explain those modifications to you later in this chapter, though. First let's look at a nuvexport session so you know what you're going to get at the end of this installation process. This is important given that the rest of MythTV is highly graphical; suddenly you have an application that doesn't have a user interface like the rest of MythTV.

Unlike the mythtranscode command discussed in the previous section, nuvexport is intended to be used as a fully interactive user interface. It is, however, also capable of being used in user jobs, which is discussed in the next section.

We don't find the lack of a TV-friendly user interface to be a big problem, though, because transcoding video is a slow process best done on something other than your television. For example, a typical movie takes several hours to transcode to the right format for an iPod video. Exactly how slow transcoding is will depend on your hardware, what quality options you select, and how busy the machine is performing other tasks. Generally we run our nuvexport sessions inside GNU screen, a text session multiplexer, so that they don't end when our ssh connection terminates. We'll show you how to do that as well.

A Typical Session

nuvexport has a text-based user interface for exporting videos from MythTV. You run a simple
command line, and it presents a menu. Here is a quick walk-through of a typical session:

```
$ nuvexport
Using ffmpeg for exporting.
What would you like to do?

 1. Export to XviD
 2. Export to SVCD
 3. Export to VCD
 4. Export to DVCD (VCD with 48kHz audio for making DVDs)
 5. Export to DVD
 6. Export to DivX
 7. Export to ASF
 8. Export to MP3
 9. Export to PSP
10. Export to iPod
11. Export to .nuv and .sql

 q. Quit

Choose a function:
```

You can see from the menu that a bunch of different export options are supported. As an
example, we'll walk through transcoding some files to the format used by a Sony PlayStation
Portable. At the initial menu, you press 9 and end up at this menu (an ellipsis indicates text
we've removed to save space and make the output more readable; text we entered is in bold):

```
You have recorded the following shows:

 1. 12 Monkeys (1 episode)
 2. 28 Days Later (1 episode)
...
35. MythBusters (13 episodes)
...
58. Whose Line Is It Anyway? (2 episodes)
59. Windtalkers (1 episode)
60. XXX (1 episode)

 q. Quit

Choose a show:  35
```

Then you'll see a list of the episodes of show 35. Here you can select a single show with just one number, a range of shows with a hyphen between two numbers, or a list of more than one show with various numbers with spaces between them (again, an ellipsis indicates skipped text, and bold indicates text we entered):

```
You have recorded the following episodes of MythBusters:

    1. Exploding Jawbreaker (3/23, 01:00 AM) 720x480 DIVX (4:3)
       Inflicting bodily harm with a playing card; heated jawbreaker.
    2. Barrel of Bricks (3/25, 7:00 PM) 720x480 DIVX (4:3)
       Hosts test urban legends: third-rail danger; electric-eel-skin wallet;
       unlucky construction-worker woes.
...
   11. Poppy Seed Drug Test (4/6, 1:00 PM) 720x480 MPEG2 (4:3)
       Hosts test urban legends: lawn-chair balloon; poppy seed bagels and drug
       tests; being covered in gold paint.
   12. Cooling a Six-Pack (4/9, 12:00 AM) 720x480 MPEG2 (4:3)
       Ancient batteries; rebuilding a crash test dummy; fastest way to cool a
       six-pack of beer.
   13. Steel Toe Amputation (4/9, 1:00 PM) 720x480 MPEG2 (4:3)
       The effectiveness of steel-toe boots; Newton's third law.

* Separate multiple episodes with spaces, ranges with '-', or * for all shows.

    r. Return to shows menu
    q. Quit

Choose a function, or desired episode(s):  1 12
```

Everything is confirmed, and you're asked whether you'd like to pick more shows. You do want to do this:

```
You have chosen to export 2 episodes:

    1. MythBusters:
       Exploding Jawbreaker (3/23, 01:00 AM) 720x480 DIVX (4:3)
       Inflicting bodily harm with a playing card; heated jawbreaker.
    2. MythBusters:
       Cooling a Six-Pack (4/9, 12:00 AM) 720x480 MPEG2 (4:3)
       Ancient batteries; rebuilding a crash test dummy; fastest way to cool a
       six-pack of beer.

* Separate multiple episodes with spaces, or ranges with '-'

    c. Continue
    n. Choose another show
    q. Quit

Choose a function, or episode(s) to remove:  n
```

You're taken back to the main menu from the previous listing. You should note that if you picked all the episodes of a given show, then that show is dropped from the listing that is offered to you. Let's export a movie as well:

```
You have recorded the following shows:

  1. 12 Monkeys (1 episode)
  2. 28 Days Later (1 episode)
...
 35. MythBusters (13 episodes)
...
 58. Whose Line Is It Anyway? (2 episodes)
 59. Windtalkers (1 episode)
 60. XXX (1 episode)

 q. Quit

Choose a show:  2
```

Again, what you have selected so far is confirmed. Note that you can also remove shows you regret selecting here. Start transcoding now:

```
You have chosen to export 3 episodes:

 1. MythBusters:
    Exploding Jawbreaker (3/23, 01:00 AM) 720x480 DIVX (4:3)
    Inflicting bodily harm with a playing card; heated jawbreaker.
 2. MythBusters:
    Cooling a Six-Pack (4/9, 12:00 AM) 720x480 MPEG2 (4:3)
    Ancient batteries; rebuilding a crash test dummy; fastest way to cool a
    six-pack of beer.
 3. 28 Days Later:
    (3/20, 3:30 PM) 720x480 DIVX (4:3)
    Survivors (Cillian Murphy, Noah Huntley, Naomie Harris) try to stay a step
    ahead of vicious, virus-infected humans that have overrun London.

* Separate multiple episodes with spaces, or ranges with '-'

 c. Continue
 n. Choose another show
 q. Quit

Choose a function, or episode(s) to remove:  c
Where would you like to export the files to? [.] /data/psp
Enable Myth cutlist? [Yes] <return to accept the default>
Enable noise reduction (slower, but better results)? ➡
[Yes] <return to accept the default>
Enable deinterlacing? [Yes] <return to accept the default>
Crop broadcast overscan (2% border)? [Yes] <return to accept the default>
```

Frame rate (high=29.97, low=14.985)? [low] **<return to accept the default>**
Resolution (320x240, 368x208, 400x192)? [320x240] **<return to accept the default>**
Video bitrate (high=768, low=384)? [high] **<return to accept the default>**
Create thumbnail for video? [Yes] **<return to accept the default>**

Finally, the transcode starts. This can take a long time, as shown by the output of the transcoding job we just selected, which ran on a 2.8GB hyperthreaded Pentium 4. Here's how the output for one of the episodes looks:

```
In order for the movie files to be of use on your PSP, you must
copy the movie file (and thumbnail if present) to the PSP's memory
stick in the following location (create the directories if necessary):

  \mp_root\100mnv01\

The movie must be renamed into the format M4V<5 digit number>.MP4.
If you have a thumbnail, it should be named in the same way, but have
a THM file extension

Now encoding:  MythBusters:  Cooling a Six-Pack
Encode started:  Tue Apr 11 19:34:43 2006
Waiting for mythtranscode to set up the fifos.
Waiting for mythtranscode to set up the fifos.
Starting ffmpeg: /usr/bin/nice -n19 ffmpeg -f rawvideo -s 720x480 -r 29.970 ➡
-i /tmp/fifodir_9887/vidout -f yuv4mpegpipe - 2> /dev/null | /usr/bin/nice -n19 ➡
yuvdenoise 2> /dev/null | /usr/bin/nice -n19 ffmpeg -threads 2 -y -f s16le ➡
-ar 48000 -ac 2 -i /tmp/fifodir_9887/audout -f yuv4mpegpipe -s 720x480 ➡
-aspect 1.33333333333333 -r 29.970 -i - -aspect 1.3333 -r 14.985 ➡
-deinterlace -croptop 10 -cropbottom 10 -cropleft 14 -cropright 14 -s 320x240 ➡
-b 768 -bufsize 65535 -ab 32 -acodec aac -f psp ➡
-title 'MythBusters - Cooling a Six-Pack' -ar 24000 ➡
'/data/psp/MythBusters - Cooling a Six- Pack.MP4'
processed:  53886 of 107805 frames (49.98%),   3.18 fps
mythtranscode finished.
processed:  53888 of 107805 frames (49.99%),   3.18 fps
ffmpeg finished.
ffmpeg version CVS, build 3276800, Copyright (c) 2000-2004 Fabrice Bellard
  configuration:  --extra-cflags=-fomit-frame-pointer -DRUNTIME_CPUDETECT ➡
--build i486-linux-gnu --enable-gpl --enable-pp --enable-zlib --enable-vorbis ➡
--enable-libogg --enable-theora --enable-a52 --enable-dts --enable-dc1394 ➡
--enable-libgsm --disable-debug --enable-mp3lame --enable-faad --enable-faac ➡
--enable-xvid --prefix=/usr
  built on Apr  3 2006 15:42:27, gcc: 4.0.3 (Ubuntu 4.0.3-1ubuntu3)

Seems that stream 0 comes from film source: 2997.00 (2997/1) -> 15.00 (15/1)
Input #0, mov,mp4,m4a,3gp,3g2, from ➡
'/data/psp/MythBusters - Cooling a Six- Pack.MP4':
```

```
Duration: 00:59:56.1, start: 0.000000, bitrate: 758 kb/s
  Stream #0.0: Video: mpeg4, yuv420p, 320x240, 2997.00 fps
  Stream #0.1: Audio: aac, 24000 Hz, stereo
Output #0, mjpeg, to '/data/psp/MythBusters - Cooling a Six- Pack.THM':
  Stream #0.0: Video: mjpeg, yuvj420p, 160x120, 1.00 fps, q=2-31, 200 kb/s
Stream mapping:
  Stream #0.0 -> #0.0
[mjpeg @ 0x83b86c8]removing common factors from framerate
Press [q] to stop encoding
frame=    1 q=0.0 Lsize=       3kB time=1.0 bitrate=  26.7kbits/s
video:3kB audio:0kB global headers:0kB muxing overhead 0.000000%

Encode finished:  Wed Apr 12 00:17:28 2006
Encode lasted: 4h 42m 45s
```

And here is the output for the movie:

```
Now encoding:  28 Days Later:  Untitled
Encode started:  Tue Apr 11 04:57:34 2006
Waiting for mythtranscode to set up the fifos.
Waiting for mythtranscode to set up the fifos.
Starting ffmpeg: /usr/bin/nice -n19 ffmpeg -f rawvideo -s 720x480 ➡
-r 29.9700298309326 -i /tmp/fifodir_9295/vidout -f yuv4mpegpipe - 2> ➡
/dev/null | /usr/bin/nice -n19 yuvdenoise 2> /dev/null | /usr/bin/nice -n19 ➡
ffmpeg -threads 2 -y -f s16le -ar 48000 -ac 2 -i /tmp/fifodir_9295/audout ➡
-f yuv4mpegpipe -s 720x480 -aspect 1.33333333333333 ➡
-r 29.9700298309326 -i - -aspect 1.3333 -r 14.985 -deinterlace -croptop 10 ➡
-cropbottom 10 -cropleft 14 -cropright 14 -s 320x240  -b 768 -bufsize 65535 ➡
-ab 32 -acodec aac -f psp -title '28 Days Later - Untitled' -ar 24000 ➡
'/data/psp/28 Days Later.MP4'
processed:  134346 of 269610 frames (49.83%),   3.01 fps
mythtranscode finished.
processed:  134807 of 269610 frames (50.00%),   2.96 fps
ffmpeg finished.
ffmpeg version CVS, build 3276800, Copyright (c) 2000-2004 Fabrice Bellard
  configuration:  --extra-cflags=-fomit-frame-pointer -DRUNTIME_CPUDETECT ➡
--build i486-linux-gnu --enable-gpl --enable-pp --enable-zlib --enable-vorbis ➡
--enable-libogg --enable-theora --enable-a52 --enable-dts --enable-dc1394 ➡
--enable-libgsm --disable-debug --enable-mp3lame --enable-faad --enable-faac ➡
--enable-xvid --prefix=/usr
  built on Apr  3 2006 15:42:27, gcc: 4.0.3 (Ubuntu 4.0.3-1ubuntu3)

Seems that stream 0 comes from film source: 2997.00 (2997/1) -> 15.00 (15/1)
Input #0, mov,mp4,m4a,3gp,3g2, from '/data/psp/28 Days Later.MP4':
  Duration: 02:29:56.1, start: 0.000000, bitrate: 525 kb/s
  Stream #0.0: Video: mpeg4, yuv420p, 320x240, 2997.00 fps
  Stream #0.1: Audio: aac, 24000 Hz, stereo
```

```
Output #0, mjpeg, to '/data/psp/28 Days Later.THM':
  Stream #0.0: Video: mjpeg, yuvj420p, 160x120, 1.00 fps, q=2-31, 200 kb/s
Stream mapping:
  Stream #0.0 -> #0.0
[mjpeg @ 0x83b86c8]removing common factors from framerate
Press [q] to stop encoding
frame=    1 q=0.0 Lsize=       5kB time=1.0 bitrate=  44.6kbits/s
video:5kB audio:0kB global headers:0kB muxing overhead 0.000000%

Encode finished:  Tue Apr 11 17:37:30 2006
Encode lasted: 12h 39m 56s
```

Now the transcoding is finished.

Having shown you what nuvexport does and how it looks, now it's time to get nuvexport installed and give it a go for real.

Shared Dependencies

nuvexport shares some dependencies with other transcoding applications mentioned in later chapters, so I will discuss those now instead of mentioning them twice in the book. The following packages are needed for both nuvexport and mytharchive. Let's start with the easy bit, which is installing the dependencies that are packaged by Ubuntu:

```
$ sudo apt-get install libdate-manip-perl transcode lame mencoder-586 ➥
libxvidcore4-dev libfaac-dev
```

Note that some of these need the Universe and Multiverse repositories enabled on Ubuntu Dapper to be available for install. Universe and Multiverse are separate groups of packages not explicitly provided by the core Ubuntu developers. These packages are generally less tested than the core Ubuntu packages and definitely less supported by Canonical, the company behind Ubuntu. Then again, they're as well tested and supported as many other Linux distributions, so they are perfectly safe to include on your system. You can find an online utility that will help you generate an /etc/apt/sources.list file, which includes Universe and Multiverse packages at http://www.ubuntulinux.nl/source-o-matic.

You might already have some of these installed, depending on what other packages you have installed on your machine. Next, you need a newer version of the ffmpeg package than what is packaged with Ubuntu Dapper. First download the build dependencies:

```
$ sudo apt-get build-dep ffmpeg
```

Now you need to download the source code for ffmpeg. For this, you'll need to have Subversion installed, which is as easy as the following:

```
$ apt-get install subversion
```

And now you can download the code, which will produce output like this:

```
$ svn checkout svn://svn.mplayerhq.hu/ffmpeg/trunk ffmpeg
A    ffmpeg/configure
A    ffmpeg/Doxyfile
A    ffmpeg/ffmpeg.c
...
A    ffmpeg/ffserver.h
A    ffmpeg/MAINTAINERS
 U   ffmpeg
Checked out revision 5780.
```

You can now compile ffmpeg like this:

```
$ cd ffmpeg
$ ./configure --enable-gpl --enable-xvid --enable-faac
install prefix    /usr/local
source path       /tmp/ffmpeg/ffmpeg
C compiler        gcc
make              make
CPU               x86 (generic)
big-endian        no
...
network support      yes
IPv6 support         yes
License: LGPL
Creating config.mak and config.h...
$ make
make -C libavutil    all
make[1]: Entering directory `/tmp/ffmpeg/ffmpeg/libavutil'
gcc -O3  -g -Wdeclaration-after-statement -Wall -Wno-switch -DHAVE_AV_CONFIG_H -
DBUILD_AVUTIL -I.. -D_FILE_OFFSET_BITS=64 -D_LARGEFILE_SOURCE -D_ISOC9X_SOURCE
-c -o mathematics.o mathematics.c
...
$ sudo make install
Password:
make -C libavutil    all
make[1]: Entering directory `/tmp/ffmpeg/ffmpeg/libavutil'
...
```

And finally, you need to install mjpegtools from source. Like ffmpeg, there is a version of mjpegtools packaged for Ubuntu, but we found it to be buggy. If you're running an Ubuntu version newer than Dapper, then you might find that the packaged version works for you. Download the source code for mjpegtools by going to http://sourceforge.net/project/showfiles.php?group_id=5776 and clicking the Download link.

You'll be presented with some download options. Select the mjpegtools source.

Now, you need to install the build dependencies, extract the source code, and compile it as usual:

```
$ sudo apt-get build-dep mjpegtools
Password:
Reading package lists... Done
Building dependency tree... Done
Note, selecting libdv4-dev instead of libdv-dev
The following NEW packages will be installed:
  debconf-utils debhelper dpatch html2text intltool-debian libaa1-dev
  libartsc0-dev libasound2-dev libaudiofile-dev libdv4-dev libesd0-dev
  libglib1.2-dev libgtk1.2-dev libpopt-dev libquicktime-dev libraw1394-dev
  libsdl1.2-dev libslang2-dev nasm po-debconf
0 upgraded, 20 newly installed, 0 to remove and 5 not upgraded.
Need to get 6045kB of archives.
After unpacking 18.3MB of additional disk space will be used.
Do you want to continue [Y/n]? y
Get:1 http://archive.ubuntu.com dapper/main libglib1.2-dev 1.2.10-10.1build1 [157kB]
Get:2 http://archive.ubuntu.com dapper/main libgtk1.2-dev 1.2.10-18 [1147kB]
Get:3 http://archive.ubuntu.com dapper/main debconf-utils 1.4.72ubuntu9 [30.9kB]
Get:4 http://archive.ubuntu.com dapper/main html2text 1.3.2a-3 [95.5kB]
Get:5 http://archive.ubuntu.com dapper/main intltool-debian ➥
0.34.1+20050828ubuntu1 [27.7kB]
Get:6 http://archive.ubuntu.com dapper/main po-debconf 0.9.2 [103kB]
...
```

When that's done, extract the source code you downloaded, and get on with the compile:

```
$ tar xzf mjpegtools-1.8.0.tar.gz
$ cd mjpegtools-1.8.0
$ ./configure
checking build system type... i686-pc-linux-gnu
checking host system type... i686-pc-linux-gnu
checking target system type... i686-pc-linux-gnu
...
$ make
make  all-recursive
...
mkdir .libs
 gcc -DHAVE_CONFIG_H -I. -I. -I../.. -I../.. -I../../utils -march=pentium3 -mtun
e=pentium3 -g -O2 -pthread -Wall -Wunused -MT build_sub22_mests.lo -MD -MP -MF .
deps/build_sub22_mests.Tpo -c build_sub22_mests.c  -fPIC -DPIC -o .libs/build_su
b22_mests.o
...
$ sudo make install
Password:
make[1]: Entering directory `/home/mikal/Desktop/mjpegtools-1.8.0/utils'
Making install in mmxsse
...
```

mjpegtools is now ready to use, so let's move on.

Nuvexport-Specific Dependencies

Now you're ready to install the dependencies needed for nuvexport that weren't installed earlier in the "Shared Dependencies" section. The dependencies for nuvexport are rather simple now that you have the common dependencies installed from the previous section of this chapter. You should just need to install mplayer and the ID3 tag support utility. You do that with this simple command line:

```
$ sudo apt-get install libid3-3.8.3-dev mplayer
```

WHAT ARE ID3 TAGS?

ID3 tags are little nuggets of information stored inside an MP3 audio file. These were not part of the original MP3 specification but were developed independently later. These tags often include information such as the title of the track, the artist, and so forth.

You can read more about ID3 tags on the Wikipedia page at http://en.wikipedia.org/wiki/Id3 and on the official id3 page at http://www.id3.org/.

Getting nuvexport Working

Getting nuvexport working was a little bit challenging. We had to tweak the Perl code for nuvexport to get it working. The patch we developed performs two tasks. First, it lets you know what dependencies are missing to make a particular piece of functionality work; for example, if you're missing the ID3 tag support utility, then you get a message to that effect. Second, it works with newer versions of ffmpeg, which list their capabilities in a slightly different manner than what nuvexport expects.

Here is the patch:

```
Index: export/ffmpeg.pm
===================================================================
--- export/ffmpeg.pm                (revision 271)
+++ export/ffmpeg.pm                (working copy)
@@ -300,7 +300,7 @@
            $children{$cat_pid} = 'audio dump' if ($cat_pid);
        }
    # Execute ffmpeg
-       print "Starting ffmpeg.\n" unless ($DEBUG);
+       print "Starting ffmpeg: $ffmpeg\n" unless ($DEBUG);
        ($ffmpeg_pid, $ffmpeg_h) = fork_command("$ffmpeg>&1");
        $children{$ffmpeg_pid} = 'ffmpeg' if ($ffmpeg_pid);

Index: export/ffmpeg/PSP.pm
===================================================================
```

```
--- export/ffmpeg/PSP.pm    (revision 271)
+++ export/ffmpeg/PSP.pm    (working copy)
@@ -170,7 +170,8 @@
        $self->{'ffmpeg_xtra'}  = ' -b ' . $self->{'v_bitrate'}
                                 .' -bufsize 65535'
                                 .' -ab 32 -acodec aac'
-                                ." -f psp -title $safe_title";
+                                ." -f psp -title $safe_title"
+                                .' -ar 24000';
     # Execute the parent method
        $self->SUPER::export($episode, '.MP4');

Index: nuv_export/shared_utils.pm
====================================================================
--- nuv_export/shared_utils.pm              (revision 271)
+++ nuv_export/shared_utils.pm              (working copy)
@@ -66,7 +66,11 @@
        $termios->getattr;
        $OSPEED = $termios->getospeed;
    };
-    our $terminal = Term::Cap->Tgetent({OSPEED=>$OSPEED});
+
+    my $terminal = Null; # This is the controlling terminal
+    eval {
+        $terminal = Term::Cap->Tgetent({OSPEED=>$OSPEED});
+    };

 # Gather info about how many cpu's this machine has
    if (-e '/proc/cpuinfo') {
@@ -85,7 +89,15 @@

 # Clear the screen
    sub clear {
-        print $DEBUG ? "\n" : $terminal->Tputs('cl');
+        if ($DEBUG) {
+            print "\n";
+        }
+        elsif ($terminal ne Null ) {
+            $terminal->Tputs('cl');
+        }
+        else {
+            print "\n";
+        }
    }

 # Byte swap a 32-bit number from little-endian to big-endian
Index: nuv_export/ui.pm
====================================================================
```

```
--- nuv_export/ui.pm                    (revision 271)
+++ nuv_export/ui.pm                 (working copy)
@@ -380,6 +380,7 @@
            $count++;
            $query .= (' ' x (3 - length($count)))."$count. ".$exporter->{'name'};
            $query .= ' (disabled)' unless ($exporter->{'enabled'});
+           $query .= "\n\t\tErrors: @{$exporter->{'errors'}}\n" unless ➥
(exporter->{'enabled'});
            $query .= "\n";
        }
        $query .= "\n  q. Quit\n\nChoose a function: ";
```

You can download the latest version of this nuvexport patch from http://www.stillhq.com/ mythtv/. You apply the patch by taking the nuvexport code from https://svn.forevermore.net/ nuvexport/ and applying the patch with the patch command. In the following example, we use the Subversion source code control system to download the latest version of the code and apply the patch we downloaded from http://www.stillhq.com/mythtv:

```
$ svn co https://svn.forevermore.net/svn/nuvexport/trunk nuvexport
A    nuvexport/nuvexportrc
A    nuvexport/export
...
 U   nuvexport
Checked out revision 271.
$ wget http://www.stillhq.com/mythtv/nuvexport.002.patch
--21:37:03--  http://www.stillhq.com/mythtv/nuvexport.002.patch
           => `nuvexport.001.patch'
Resolving www.stillhq.com... 210.18.204.2
Connecting to www.stillhq.com|210.18.204.2|:80... connected.
HTTP request sent, awaiting response... 200 OK
Length: 3,319 (3.2K) [text/plain]

100%[===============================================>] 3,319        --.--K/s

21:37:04 (237.24 KB/s) - `nuvexport.002.patch' saved [3319/3319]
```

```
$ cd nuvexport
$ patch -p 0 < ../nuvexport.002.patch
```

Because nuvexport is a Perl program, once the patch is applied, you're ready to run. You can install nuvexport into /usr/local/bin/ manually if you want, but we haven't bothered. We instead installed nuvexport into a different location and then created a symlink from /usr/local/ bin/nuvexport to the location of the nuvexport script in the nuvexport directory and a symlink in the MythTV user's home directory from .nuvexportrc to the nuvexportrc in the nuvexport directory.

Transcoding Automatically to Other Devices

You can also have MythTV transcode videos either automatically or triggered through the user interface. The easiest way to do this is to have MythTV run nuvexport from the command line. We've already talked about how to run mythtranscode interactively from the terminal window, so we'll next explore how to get nuvexport working without any interaction except for the command line, and then it will be easy to plug both into MythTV.

nuvexportrc

You first need to modify the nuvexportrc file to provide a reasonable default for where to write the transcoded files. Your nuvexportrc file can be in the current working directory (with the name nuvexportrc), in the home directory of the user who will be running the jobs (with the name .nuvexportrc), or in /etc/ (with the name nuvexportrc). We recommend you put it in the home directory of the user who will be running the transcode jobs, because that way it won't be lost when you reinstall the system. We'll briefly cover the configuration file, even though you won't change it all so you can get a feel for the other tasks you can do within the file. Here is our nuvexportrc file, with the value you need to set in bold:

```
# Generic options that don't relate specifically to any particular exporter.
<nuvexport>

# Set export_prog to ffmpeg, transcode, or mencoder, depending on your
#   preference of program for exports.  This is equivalent to --ffmpeg,
#   --transcode, or --mencoder
#

   export_prog=ffmpeg

# You can specify a default mode, which will mean you don't need to use
# the --mode command-line option.
#
#   mode=xvid

# Setting underscores to yes will convert whitespace in filenames to an
#   underscore character (which some people seem to prefer)
#

   underscores=no

# Setting require_cutlist to yes will tell nuvexport to show only those
#   recordings that have a cutlist
#
#   require_cutlist=no
```

```
#  By default, nuvexport picks what it thinks is a good name for your file
#     (doing its best to avoid printing "Untitled" into the filename).  Setting
#     name will let you change the output format of the filename generated by
#     nuvexport.  Even after this formatting, nuvexport will still do some basic
#     replacements to make sure that illegal filename characters (eg. /\:*?<>|)
#     are replaced with a dash (or " with a ').  The following format variables
#     are supported:
#
#     %f -> full path to the filename
#     %c -> the chanid of the show
#     %a -> start time in YYYYMMDDHHMMSS format
#     %b -> end time in YYYYMMDDHHMMSS format
#     %t -> title (show name)
#     %s -> subtitle (episode name)
#     %h -> hostname where the file resides
#     %m -> showtime in human-readable format (see --date below)
#     %d -> description
#     %% -> a % character
#
#     filename=%t - %s


#  By default, nuvexport uses an American-style date to represent show times in
#     lists and filenames.  Use --date to override that with the format of your
#     choosing.  See the UnixDate section `perldoc Date::Manip` for formatting
#     options.
#
#     date=%m/%d, %i:%M %p


#  Nuvexport has the option to crop a percentage of the border of each recording
#  in order to get rid of the unsightly edges of the TV signal.  The default 2%
#  approximates the overscan of an average TV, but you can alter this from 0 to
#  5% to fit your preferences.
#
    overscan_pct = 2


</nuvexport>

<generic>

# Default to export to the current directory. You have to set this to not be
# prompted for command-line transcoding.
#
    path = /data/psp/
```

```
# Use the cutlist (not to be confused with the commercial flag list) when
#    exporting.
#

    use_cutlist = yes


# Tell mythcommflag to generate a cutlist from the commercial flags before
#    exporting.  Don't forget to enable use_cutlist above, too.
#
#   gencutlist = no


# Contrary to popular belief, enabling multipass will not make your recordings
#    look better.  What it will do, however is guarantee that the bit rate you
#    choose will be the average bit rate of your entire encode (meaning that your
#    exports will end up being about the same size per-minute) and that you
#    will receive the best overall quality for a files of the same size.
#

    multipass = yes


# Disabling noise reduction can speed up your exports dramatically, but at the
#    expense of some quality.  You can also access this on the command line via
#    the --denoise (or --nodenoise) flag.
#

    noise_reduction = yes


# Deinterlace the video so that it looks better on software players.
#

    deinterlace    = yes


# Crop about 2% from the border of the recording before encoding.  This is done
#    to get rid of part of the broadcast signal that is usually obscured by the
#    TV's overscan.
#

    crop = yes


#  If you have a particularly dirty signal, you might want to try to disable
#    fast_denoise (it's actually part of yuvdenoise, which both the ffmpeg
#    and transcode exporters call).  It can be almost twice as slow as the
#    default "fast" normal noise reduction, but it considerably more effective.
#    The latest version of yuvdenoise (which is called directly by the ffmpeg
#    exporters) does not support this option, so it is ignored in that case.
#

    fast_denoise = yes
```

```
#  If nuvexport is having trouble detecting the *input* aspect ratio of your
#     recordings (MythTV used to hard-code all software-encoded files as 1:1
#     regardless of the true aspect), set this option to one of the following:
#
#    force_aspect = [ 1:1 4:3 16:9 2.21:1 ]

</generic>


#  These are options for the different encoders

<ffmpeg>
#  ffmpeg is almost twice as fast if you disable noise reduction
#
#    noise_reduction = no

#  By default, nuvexport's ffmpeg module lets ffmpeg handle deinterlacing.
#     We've found that this provides the best results, but if you wish to let
#     yuvdenoise do it instead, set deint_in_yuvdenoise to a true value.
#
#    deint_in_yuvdenoise = no
</ffmpeg>

<transcode>
#  Mythtranscode will always be used for nupplevideo recordings because
#     transcode can't read them, but setting force_mythtranscode to yes will
#     force nuvexport to call mythtranscode when using the transcode exporter for
#     mpeg recordings, too.  This may help problems that some people have been
#     having with transcode not recognizing certain dvb recordings, as well as
#     transcode not working properly on certain ivtv recordings.
#
    force_mythtranscode = yes

#  Setting both force_mythtranscode and mythtranscode_cutlist to yes will tell
#     nuvexport to use mythtranscode's built-in cutlist functions, rather than
#     having transcode use its own.  We've found that the cutlists for a handful
#     of ivtv recordings that do not work properly with transcode's internal
#     cutlist handler.
#
    mythtranscode_cutlist = yes

</transcode>

<mencoder>
</mencoder>
```

```
#  You can also specify options for output formats

<XviD>
    vbr            = yes   # Enable vbr to get the multipass/quantisation options
                           # (enabling multipass or quantisation automatically
                           # enables vbr)
    multipass      = yes   # You get either multipass or quantisation; multipass
                           # will override
    quantisation = 6       # 4 through 6 is probably right...  1..31 are allowed
                           # (lower is better quality)

    a_bitrate      = 128   # Audio bitrate of 128 kbps
    v_bitrate      = 960   # Remember, quantisation overrides video bitrate

    width          = 624   # Height adjusts automatically to width, according to
                           # aspect ratio
    height         = auto

</XviD>

<MP3>
#  Default mp3 bitrate is 128
#
    bitrate = 128
</MP3>

#  If you want to provide settings for a very specific export module, you can
#    use its full name, and it will override any more generic settings.

<ffmpeg::PSP>
# PSP framerate (high=29.97, low=14.985)
    psp_fps = low

# PSP resolution (320x240, 368x208 or 400x192)
    psp_resolution = 320x240

# PSP video bitrate (high=768, low=384)
    psp_bitrate = high

# Create a thumbnail to go with the PSP video export?
    psp_thumbnail = yes
</ffmpeg::PSP>

<mencoder::XviD>
    multipass = no
</mencoder::XviD>
```

The only setting you really need to change from the default is the one we have in bold, which is the output path for the videos.

nuvexport Command Line

nuvexport supports a command-line interface as well as the interactive text interface. This means you can eliminate almost all the interaction at transcode time, which is vital for automating transcode jobs with MythTV. The command-line options were poorly documented at the time of writing this chapter, but trawling the source code has revealed the following options.

WHY IS /VAR/VIDEO REPEATED?

The repetition of /var/video in some filename lines in the following sections is because of a nuvexport bug. We store videos in /var/video, and that has been prepended to the string incorrectly.

--help

This displays a help message for the command line. This isn't very helpful at the moment, though, because it refers you to a web page.

--debug

This displays debugging output.

--version

This displays the version of nuvexport and then exits.

--nice

Transcoding is CPU intensive, and it is likely that you will want to specify the priority to run the transcode job at so that the machine still remains responsive for more important tasks such as video playback if this is a frontend machine. The --nice option specifies the nice level as per the nice command. To learn more, run this from a terminal:

```
man nice
```

--mode

This specifies the output format for the video. Possible values are as follows:

- xvid
- svcd
- vcd
- dvcd
- dvd
- divx
- asf
- mp3
- psp
- ipod

The names of these modes are not case-sensitive. These modes depend on you having the right libraries installed to make transcoding for that mode work, though. The rest of the examples in the following sections will use the psp output mode.

--title

This specifies the title of the video to transcode. This needs to be exactly the title as displayed in the MythTV user interface. For example:

```
nuexport --title "JoJo's Circus"
```

--subtitle

The title by itself is almost certainly not unique enough to identify a particular video. For example, *JoJo's Circus* is a series, and you need to specify an episode to transcode. Again, an exact text match is needed here:

```
nuvexport --title "JoJo's Circus" --subtitle "Funny Bunnies; Ivan the Bearable"
```

--episode is a synonym for --title.

--description

Alternatively, you can transcode shows based on their descriptions. For example, let's say you are particularly interested in the superheroes game from "Whose Line Is It Anyway?" You might start with the following command line:

```
$ nuvexport --description Superheroes --mode psp -search-only
Loading MythTV recording info.
99%

Matching Shows:

    title:  The Backyardigans
 subtitle:  Race to the Tower of Power
   chanid:  1005
   starts:  20060715083000
     ends:  20060715090000
 filename:  /var/video//var/video/1005_20060715083000.mpg

    title:  Whose Line Is It Anyway?
 subtitle:  Untitled
   chanid:  1052
   starts:  20060724223000
     ends:  20060724230000
 filename:  /var/video//var/video/1052_20060724223000.mpg

    title:  Whose Line Is It Anyway?
 subtitle:  Untitled
   chanid:  1052
   starts:  20060725223000
     ends:  20060725230000
 filename:  /var/video//var/video/1052_20060725223000.mpg
```

We've used the --search-only option in this command line to specify that we don't want to actually do any transcoding. You'll learn more about the search-only option soon.

That didn't work exactly how we wanted it to work, because it came up with one of the children's shows that the kids like as well. However, if you specify the title as well, you get the right list of shows:

```
$ nuvexport --title "Whose Line Is It Anyway?" --description Superheroes ➥
--mode psp--search-only
Loading MythTV recording info.
99%

Matching Shows:

    title:  Whose Line Is It Anyway?
 subtitle:  Untitled
   chanid:  1052
   starts:  20060724223000
     ends:  20060724230000
 filename:  /var/video//var/video/1052_20060724223000.mpg
```

```
     title:   Whose Line Is It Anyway?
  subtitle:   Untitled
    chanid:   1052
    starts:   20060725223000
      ends:   20060725230000
  filename:   /var/video//var/video/1052_20060725223000.mpg
```

--infile

If you know the name of the file that you want to transcode, then you can use the `--infile` argument to specify the video to transcode. For example:

```
$ nuvexport --infile /var/video/1055_20060627090000.mpg --mode psp --search-only
Loading MythTV recording info.
99%
```

```
Matching Shows:

     title:   JoJo's Circus
  subtitle:   Funny Bunnies; Ivan the Bearable
    chanid:   1055
    starts:   20060627090000
      ends:   20060627093000
  filename:   /var/video//var/video/1055_20060627090000.mpg
```

Again, we've used the `--search-only` option in this command line to stop transcoding from occurring. You'll learn more about the search-only option soon.

--chanid

You can see in the output for the previous command that the show lists a chanid, which identified the channel on which the show aired. If you want to transcode all the shows from that channel, then you can pass this chanid to the `--chanid` command-line option.

--search-only

The search-only argument will result in everything being done, except the actual transcoding. This is a good way to test command lines. For example, here is a command line where you just want to see that it matches a show and its output:

```
$ nuvexport --title "JoJo's Circus" ➡
--subtitle "Funny Bunnies; Ivan the Bearable" --mode psp -search-only
Loading MythTV recording info.
99%
```

```
Matching Shows:

      title:  JoJo's Circus
   subtitle:  Funny Bunnies; Ivan the Bearable
    chanid:   1055
    starts:   20060627090000
      ends:   20060627093000
   filename:  /var/video//var/video/1055_20060627090000.mpg
```

You can see here that the search returned only one show, so only that show would be transcoded if you removed the `--search-only`.

--confirm

`--confirm` forces all the parameters specified on the command line and in the `nuvexportrc` file to be reconfirmed by the user before the transcode actually starts:

```
$ nuvexport --title "JoJo's Circus" --subtitle "Funny Bunnies; Ivan the Bearable" ➥
--mode psp --confirm
Loading MythTV recording info.
99%

You have chosen to export 1 episode:

  1. JoJo's Circus:
     Funny Bunnies; Ivan the Bearable (6/27, 09:00 AM) 720x480 MPEG2 (4:3)
     JoJo and Skeebo watch after a magician's rabbits; JoJo and her friends
     communicate with a performing bear from Russia.

* Separate multiple episodes with spaces, or ranges with '-'

  c. Continue
  n. Choose another show
  q. Quit

Choose a function, or episode(s) to remove:  c
Where would you like to export the files to? [/data/psp]
Enable Myth cutlist? [Yes]
Enable noise reduction (slower, but better results)? [Yes]
Enable deinterlacing? [Yes]
Crop broadcast overscan (2% border)? [Yes]
Frame rate (high=29.97, low=14.985)? [low]
Resolution (320x240, 368x208, 400x192)? [320x240]
Video bitrate (high=768, low=384)? [high]
Create thumbnail for video? [Yes]
```

In order for the movie files to be of use on your PSP, you must
copy the movie file (and thumbnail if present) to the PSP's memory
stick in the following location (create the directories if necessary):

```
\mp_root\100mnv01\
```

The movie must be renamed into the format M4V<5 digit number>.MP4.
If you have a thumbnail, it should be named in the same way, but have
a THM file extension

```
Now encoding:  JoJo's Circus:  Funny Bunnies; Ivan the Bearable
Encode started:  Thu Aug  3 13:54:59 2006
Waiting for mythtranscode to set up the fifos.
Waiting for mythtranscode to set up the fifos.
Starting ffmpeg: /usr/bin/nice -n19 ffmpeg -f rawvideo -s 720x480 -r 29.970 ➡
-i /tmp/fifodir_20516/vidout -f yuv4mpegpipe - 2> /dev/null | /usr/bin/nice ➡
-n19 yuvdenoise 2> /dev/null | /usr/bin/nice -n19 ffmpeg -threads 2 -y ➡
-f s16le -ar 48000 -ac 2 -i /tmp/fifodir_20516/audout -f yuv4mpegpipe ➡
-s 720x480 -aspect 1.33333333333333 -r 29.970 -i - -aspect 1.3333 ➡
-r 14.985 -deinterlace -croptop 10 -cropbottom 10 -cropleft 14 -cropright 14 ➡
-s 320x240  -b 768 -bufsize 65535 -ab 32 -acodec aac -f psp ➡
-title 'JoJo'\''s Circus - Funny Bunnies; Ivan the Bearable' -ar 24000 ➡
'/data/psp/JoJo'\''s Circus - Funny Bunnies; Ivan the Bearable.MP4'
processed:  0 of 53835 frames (0.00%),   0.00 fps
```

You can see in this example that the episode to export had been populated by the command line, but we were then prompted to confirm everything, much like a regular interactive session.

Other Options

A few more options exist that we don't want to spend as much time covering, because they aren't as useful as those covered previously. Quickly, you can also have nuvexport transcode based on the start time of the show (--starttime), require a cut list (--require_cutlist), force an aspect ratio (--force_aspect), and decide whether nuvexport runs as a server (--noserver). The cut list option here is the most interesting, but we'll defer the discussion of cut lists to Chapter 6.

Setting Up the User Jobs

Now you're finally ready to set up the automated transcoding for MythTV recordings. We needed to cover all the previous information, because you need to have already specified the output directory for videos in the nuvexportrc file for this to work, and you need the command-line options to get nuvexport to run on the right file.

You set up automated transcoding via the user jobs, which you configure in the `mythtv-setup` application. Go to the General menu, and then hit Next until you end up at the user jobs page.

Tip You might find that you don't have a `mythtv-setup` program. That's probably because on some distributions it is named `mythtvsetup`.

Figure 5-19 shows what a default page looks like.

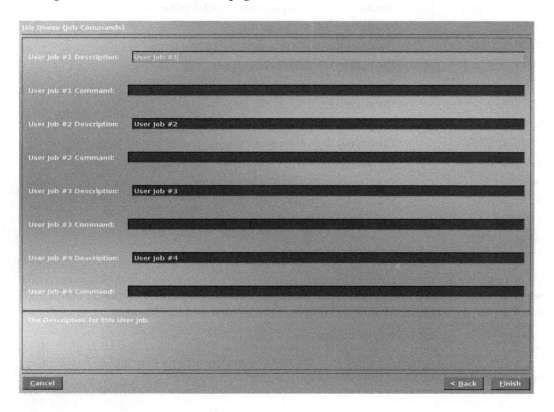

Figure 5-19. *The default user jobs configuration screen*

Your user jobs can have special variables passed through to them, and this helps you craft a `nuvexport` command line that will transcode the show correctly. The variables listed in Table 5-2 are available.

Table 5-2. *Variables Available for User Job Command Lines*

Variable	Will Contain
%CHANID%	The channel ID
%STARTTIME%	The start time of the recording
%STARTTIMEISO%	The start time of the recording in ISO date format
%ENDTIME%	The end time of the recording
%ENDTIMEISO%	The end time of the recording in ISO date format
%PROGSTART%	The start time of the program
%PROGSTARTISO%	The start time of the program in ISO date format
%PROGEND%	The end time of the program
%PROGENDISO%	The end time of the program in ISO date format
%DIR%	The path to the directory containing the video file
%FILE%	The filename of the video file
%TITLE%	The show title
%SUBTITLE%	The show subtitle
%DESCRIPTION%	The description of the show
%CATEGORY%	The category the show appears in
%RECGROUP%	The recording group used by the show
%PLAYGROUP%	The play group used by the show
%HOSTNAME%	The name of the machine running the job
%VERBOSELEVEL%	How verbose to be

This makes it easy to construct a command line that will just transcode the show currently selected in the MythTV user interface. For example, based on the earlier nuvexport command-line discussion and these variables, you can just use a command line like this:

```
nuvexport -infile %DIR%/%FILE% --mode psp
```

To transcode a particular video into the right format for a Sony PlayStation Portable, Figure 5-20 shows our actual user jobs at the moment.

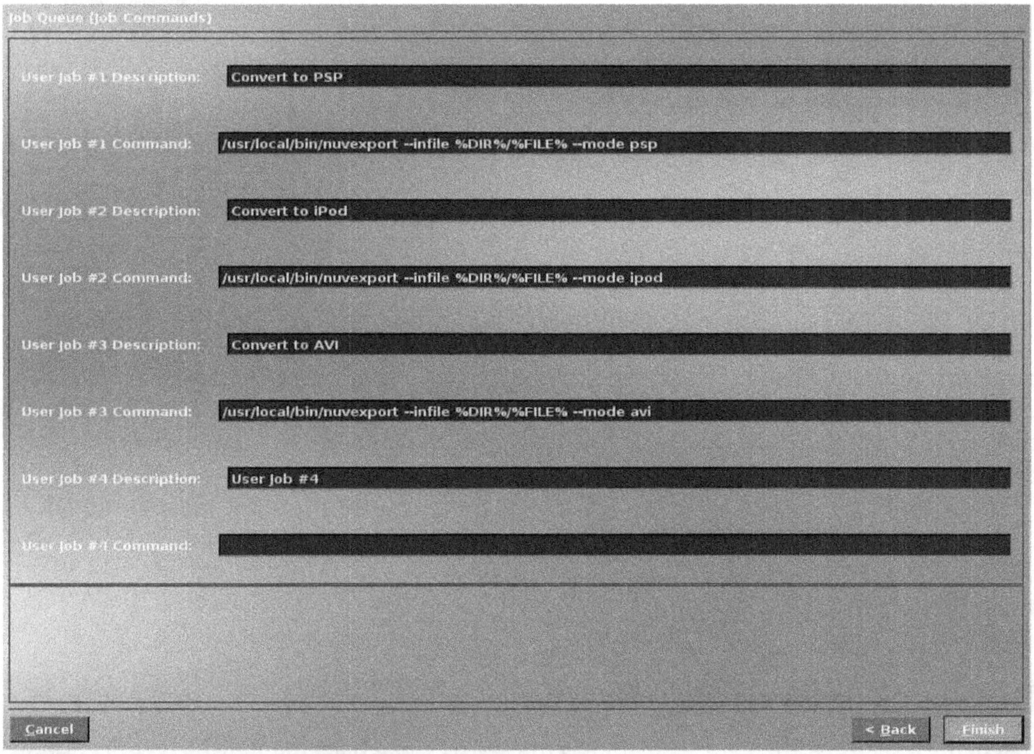

Figure 5-20. *Sample user jobs*

Conclusion

We've covered a lot of ground in this chapter, including how to set up advanced aspects of recordings, such as setting the recording options we didn't discuss in Chapter 4, automatically detecting and skipping commercials, autoexpiring recordings to save on disk space, transcoding (converting video to other formats), transcoding automatically, and adding user-defined jobs to MythTV. In the next chapter, we will discuss some of the other features offered by MythTV such as advanced aspects of recording playback before moving on to how to set up extra frontends and play DVDs in later chapters.

CHAPTER 6

■ ■ ■

Exploring MythTV Functionality

In this chapter, we'll give you an overview of the other useful MythTV features that are above and beyond the core functionality that we covered earlier in this book. MythTV offers lots of other useful features. Although we aren't able to cover all of the features in this chapter we will cover what we believe to be the most commonly used extra functionality.

Specifically, we'll cover some of the more interesting portions of the playback menu, including how to edit recordings, how to convert automatically detected commercials to cut points, how to transcode to make your edits permanent, how to skip commercials, how to set up closed captioning (also known as *subtitles*), and how to set up channel icons.

Using the Playback Menu

Most of these features are made available by the MythTV playback menu, which we'll cover in the following sections. Pressing the menu button on your remote or the M key on your keyboard during the playback of a recording or while watching Live TV gives you access to this additional functionality through the on-screen playback menu. Note that opening the menu doesn't pause playback; you'll have to do that yourself. You navigate the menu by using the arrow keys and pressing Enter to select. Figure 6-1 shows an example of the menu.

Figure 6-1. *On-screen menu during playback*

In the following sections, we'll walk you through each of the options available during playback.

The Edit Recording Option

We'll start by discussing probably the most commonly used functionality in the menu, the facility to edit recordings. Recording editing takes the form of inserting cut points into the recording. These essentially mark where you want parts of the recording removed when the recording is further processed. For example, the transcoding discussed in Chapter 5 can respect the cut points you define in the user interface and therefore produce smaller recordings that don't have extraneous video in them. It's important to note that a series of cut points is occasionally referred to as a *cut list*.

When you select Edit Recording in the menu, you'll see a bar representing the recording, as shown in Figure 6-2.

The black box is your current position in the recording. (Exactly how the bar and markers look, of course, depends on the MythTV theme you've chosen.) The left and right arrows on your keyboard or remote jump backward and forward through the show. The up and down arrows select the size of the jump used by the left and right arrow keys—anything from one frame to ten minutes. Pressing Enter on the keyboard or OK on your remote allows you to insert a cut point, deleting either before the current position or after the marker, as shown in Figure 6-3.

Figure 6-2. *The edit recording bar*

Figure 6-3. *Inserting a cut point*

To remove the beginning of the recording from this point, select Delete Before This Frame. To delete to the end of the recording from this point, select Delete After This Frame. If you want to remove a region in the middle of the recording, you can insert another cut point and use that to create a region. Create the new cut point, and then have that one delete in the direction of the first cut point. You can think After This Frame as setting the marker for the *beginning* of a region you want to cut—a commercial break, for example—and Delete Before This Frame as setting the marker for the *end* of that region.

To insert a cut point in an area that has been marked for deletion, you have to navigate in the edit recording mode; you can't play through this deleted section (one of the points of deleting a part of the recording is to have it not be played). Note that there's also a limit to how closely together you can place edit points; this is roughly 20 seconds. Figure 6-4 shows a recording that has had a commercial break removed and the start and end of the recording trimmed to the start and end of the show.

Pressing Page Up or Page Down (or Q and Z) jumps to the next or previous cut point (or the start/end of the recording). Pressing M (the menu key) will exit the edit recording screen and resume playback.

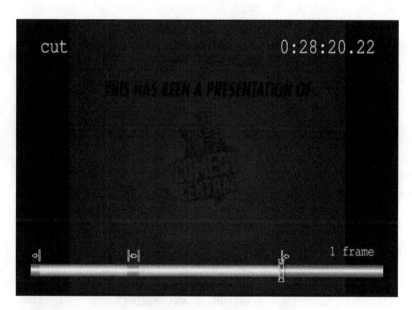

Figure 6-4. *An edited recording*

You can also edit cut points. By moving to a place close to a cut point, you can press Enter again to delete the cut point, flip the direction of the cut point, or move the cut point to the current position.

Inserting Cut Points from Commercial Detection

When on the edit recording screen, you can press the End key, Page Down key, or the key you have mapped to this function on your remote to insert all the cut points found during commercial detection. Now is a good time to review these cut points for accuracy because commercial detection isn't always flawless (and can vary from program to program and from station to station). You can also convert detected commercials to cut points from the command line like this:

```
$ mythcommflag --gencutlist -f <complete path to the video file>
```

You can create a user job to perform this command if you want. An excellent example, including a script, is available at http://mythtv.org/wiki/index.php/Removing_Commercials#Automatically_removing_commercials. Or, you may decide not to use commercial detection at all—setting the cut points for the four or five breaks in the average program, once you're practiced in it, usually takes only a minute or two. If you keep programs around, or archive to DVD, the investment in manual editing is often well worth it.

Making Your Edits Permanent

To make your edits permanent, be they from a series of cut points you created yourself or from importing a cut list from the commercial autodetection, simply transcode the recording. This can be to a different quality or to the same quality. You can find more information about the transcoding options in the playback menu in the "Begin Transcoding" section of this chapter.

A Word of Warning on Cut Points

Remember that it can be quite hard to tell whether you're on exactly the right frame when you set a cut point. Many people prefer to keep from 1 to 1.5 seconds extra per cut point when creating cut points. This way you can ensure that you definitely don't remove something you want.

Jump to Program

The Jump to Program menu gives you a quick way to move between recordings. It does this by presenting basically the same user interface as the watch recordings screen but in a much more compressed form. You'll see a submenu that lists all the titles that are currently recorded, and you can then select an episode of a title to jump to.

Begin Transcoding

During playback you can start the transcoding process. This transcoding is different from the nuvexport transcoding mentioned in Chapter 5, which is intended for exporting recordings to other formats. The transcoding available from this menu will convert to other formats supported internally by MythTV, and it is usually used for reducing the quality of a recording to save disk space or to permanently remove video marked with a cut list. It is implemented with mythtranscode, which is more fully explored in Chapter 5. You will continue watching the non-transcoded stream while transcoding is taking place. From the same menu, you can also stop the transcoding if you decide you made the wrong choice (such as if you decide to use a different transcoding profile). And, obviously, you can't start transcoding while watching the show if you're going to edit commercials manually. Figure 6-5 shows an example of the transcoding menu during playback.

Figure 6-5. *Transcoding menu during playback*

Commercial Auto-Skip

Commercial detection is one of the best features of MythTV. We discussed the setup for commercial detection in Chapter 5, so if you want to know more about commercial detection, you should check out that chapter again. You will occasionally find a recording where your default commercial skip option isn't the best one. For example, commercial skipping tends to cut out important parts of some of our *Law & Order* episodes—mainly because there are many dark frames between scene changes—but it does a great job with *Dirty Jobs*. You can change the commercial skipping option being used for a particular playback by selecting the Commercial Auto-Skip option from the playback menu. You'll then see the menu shown in Figure 6-6.

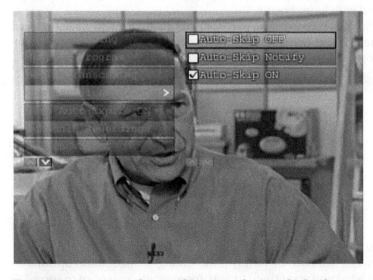

Figure 6-6. *Commercial Auto-Skip menu during playback*

The options here are as follows:

- Auto-Skip Off disables automatic commercial skipping. You can still use the detected commercial boundaries by using the left and right track-skip buttons on your remote control.

- Auto-Skip Notify lets you know when there is a commercial detected to skip but doesn't skip it. Again, you can use the track-skip buttons to jump to the closest detected commercial boundary.

- Auto-Skip On just automatically skips the commercials.

Turn Auto-Expire Off

It's quite common for us to be very far behind on shows that we're particularly keen on seeing, such as episodes of *Mythbusters*. The Turn Auto-Expire Off option allows us to instruct MythTV to not delete these episodes regardless of how little disk remains on the system. MythTV will instead delete shows that are newer than this one. If autoexpiry is off for this recording, then you can turn it back on with this menu item as well.

Change Aspect Ratio

Aspect ratio is the ratio between the width of an image and the height of the image. This is important for television programs and movies, where this aspect ratio should be as close a match as possible to your display device's aspect ratio to avoid letter boxing. *Letter boxing* is the black rectangles that are often placed at the top and bottom of the screen because the image being displayed is not tall enough.

Sometimes MythTV will detect the aspect ratio for a recording incorrectly, which is why this menu item is necessary. You can also use it to force recordings to play back at a given aspect ratio, which will either result in the image being cropped or result in it being stretched depending on which option you select. This cropping or stretching isn't bad if the image was actually at the other size or was recorded with letter boxing.

We recommend playing with this menu during playback, because it's much easier to learn how the menu works with some brief experiments than it is to describe it in writing. Briefly, the aspect ratio options are as follows:

- 4:3

- 16:9

- 4:3 Zoom

- 16:9 Zoom

- 16:9 Stretch

Manual Zoom Mode

Manual zoom mode allows you to move the image around on the screen, as well as manually zoom in on a portion of the image. This can be handy if the Change Aspect Ratio options described previously do not do what you want them to or if, for example, you want to prove that the name of the character on *Grey's Anatomy* is actually "Calliope" Torres by zooming in on her lab-coat embroidery. Table 6-1 lists the relevant keys to use while manually zooming.

Table 6-1. *Keys for Manual Zoom Mode*

Key	Function
Left	Moves the image left
Right	Moves the image right
Up	Moves the image up
Down	Moves the image down
PageUp	Zooms in
PageDown	Zooms out
Space/Enter	Exits zoom mode, leaving picture at current size and position
ESC	Exits zoom mode and returns to original size

Adjust Audio Sync

It's possible if you have a problem with your tuner card (or perhaps a network problem) for the audio and video channels of a recording to become out of sync. Nothing is more annoying than having to watch a show when the lip sync is off. MythTV of course gives you a way to tweak the synchronization between the audio and video channels; you can access it via the Adjust Audio Sync menu item.

If you select this menu item, you will see a dialog box in the top right of the video. Use the left and right arrows to adjust the audio sync.

Adjust Time Stretch

You can also choose to play a recording faster or slower than the speed at which it was broadcast. If you select this menu item, you will see a list of choices for the playback speed. These choices range from slower than normal to faster than normal, and they can be quite handy for programs where you want only an overview of what is happening (for example the daily news). It's possible to speed a show up by as much as 15 percent without appreciably affecting playback, which can be helpful if you watch a lot of TV. Alternatively, you can slow playback down quite a bit (to as little as 50 percent normal speed) to make it easier to understand the dialogue on *The West Wing* reruns.

Video Scan

This menu item lets you change the interlacing method used for the recording playback. You can choose from these options:

- Detect (the default, which will use a combination of your global settings and the video file to determine what interlacing method to use)

- Progressive

- Interlaced Normal

- Interlaced Reverse

Instead of giving a theoretical description of these interlacing options, you are much better off playing with the various options during video playback and selecting the one that looks the best for that video.

Sleep

Sleep is a lot like the sleep mode on some radios and alarm clocks. You tell MythTV how long you want the frontend to stay on, and then it will turn off after the specified time. You can use this to start a movie, music channel, or whatever else you find restful—including a playlist of previously recorded programs—to watch or listen to from bed. After the given time, the frontend will stop playback.

Closed Captioning (Subtitles)

Closed captions (also called *subtitles* or *teletext*) are sometimes broadcast in a computer-readable format and not embedded in the TV image. Most often, these are captions in the same language as what is being spoken and are aimed at those people who are hard of hearing or deaf. Many people without any hearing problems also find the closed captions useful, such as when in noisy environments. MythTV records the closed captions—if your tuner card provides for this—and you can enable (or disable) them during playback. From the menu during playback, you can select which caption set to use. The name varies depending on the delivery system you are using to get television. For example, for cable TV recordings, VBI CC is the name. Select that and then which subtitle track you want to view (Figure 6-7). MythTV will then render the subtitles on the image (Figure 6-8).

If you don't see a closed caption menu option, then it probably means you haven't configured the Vertical Blanking Interrupt (VBI) options with `mythtv-setup`. VBI is the encoding method used for PAL teletext and NTSC closed captions, and you configure it via the general options. Look for the VBI options on the third screen of the general options wizard.

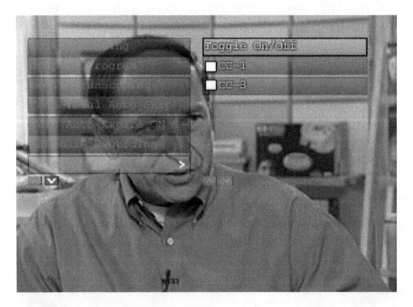

Figure 6-7. *Selecting cable TV subtitles*

Figure 6-8. *Watching cable TV subtitles*

Using Channel Icon Maps (Giving TV Stations Icons)

Depending on your source of guide data, you may already have icons being displayed for the TV stations. If not, you can easily (especially if you live in North America) add icons for your TV stations.

If you're in North America, the mkiconmap.pl script that comes with MythTV will ask you what your lineup is and produce an iconmap.xml file that you can load into MythTV. Run the mkiconmap.pl script like this (after using cd to change directory into your MythTV source directory):

```
$ perl contrib/mkiconmap.pl
```

You can now use mythfilldatabase to import the icon map into MythTV:

```
$ mythfilldatabase --import-icon-map iconmap.xml --update-icon-map
```

If you're not in North America or your service isn't listed, you'll need to construct your own iconmap.xml file. It's a rather simple file format, so don't worry. The start of the file should look like this:

```
<?xml version="1.0" encoding="UTF-8"?>
<iconmappings>
```

Then you need to specify a base URL for the images. You should download channel icons and place them in this directory. Here we've set a local folder as where we're storing the icons:

```
<baseurl>
        <stub>[local]</stub>
        <url>file:///home/myth/station-logos</url>
</baseurl>
```

You can now enter in the information for the channels. You need to make two mappings: call sign to network and network to URL. The call-sign-to-network mapping maps the call sign of the channel in your MythTV setup (see the MythWeb channel configuration page or the list in mythtv-setup) to a television network. Each network has its own logo because you may receive several channels from the same network (for example, in standard definition, high definition, +2hrs, and so on). A simple setup of mapping two channels to one network is as follows:

```
<callsigntonetwork>
        <callsign>TEN Digital</callsign>
        <network>TEN Network</network>
</callsigntonetwork>

<callsigntonetwork>
        <callsign>TEN HD</callsign>
        <network>TEN Network</network>
</callsigntonetwork>

<networktourl>
        <network>TEN Network</network>
        <url>[local]/10.tif</url>
</networktourl>
```

You now need to insert the following at the end of the file:

```
</iconmappings>
```

You can now load the new icon map into MythTV:

```
$ mythfilldatabase --import-icon-map iconmap.xml –update-icon-map
```

You can easily see the changes in the MythWeb interface and the frontend. Unfortunately, some errors in the iconmap.xml file won't be reported by mythfilldatabase (such as duplicate network-to-URL mappings). Be sure to start simple and gradually expand the list—this will help you narrow down the cause of the problem if you get something wrong.

The images should be square and will be resized to be square if they are not, sometimes with odd results. You'll easily be able to see whether you get it wrong.

Conclusion

In this chapter, we covered a few extra features and tricks of MythTV that are often useful that we hadn't already covered. Lots more features could be useful for you too—some of them you've probably already seen in menus in MythTV. There is generally no harm in experimenting with settings to see what you like and what's useful. This is exactly how we mastered MythTV, and we encourage others to do the same.

CHAPTER 7

■■■

Setting Up MythTV Themes

Like a lot of software today, MythTV is "themable." In other words, you can choose from a variety of different looks as well as create your own. The default G.A.N.T. theme can be either exciting or boring depending on your point of view. MythTV also has themes optimized for wide-screen displays.

In addition to themes, you can of course change the font size used by the frontend user interface, which will affect how the MythTV frontend looks as well. We'll discuss nontheme elements that you can configure in the "Choosing a Theme" section.

In this chapter, we will cover why you might choose some of the themes that come with MythTV, and we'll cover how to perform some simple modifications to an existing theme.

Choosing a Theme

Many people find the default G.A.N.T. theme not as exciting to look at as some of the commercial PVR interfaces. Figures 7-1 through 7-7 show the following MythTV themes that ship with MythTV:

- G.A.N.T. (see Figure 7-1)

- Blue (see Figure 7-2)

- Retro (see Figure 7-3)

- Minimalist-wide (see Figure 7-4)

- Iulius (see Figure 7-5)

- Titivillus (see Figure 7-6)

- MythCenter (see Figure 7-7)

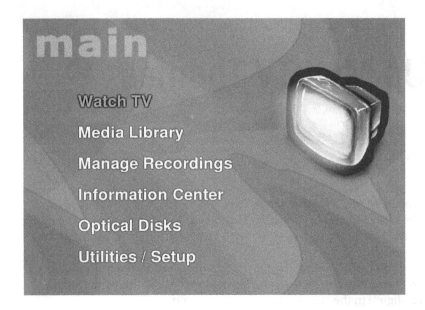

Figure 7-1. *The G.A.N.T. theme*

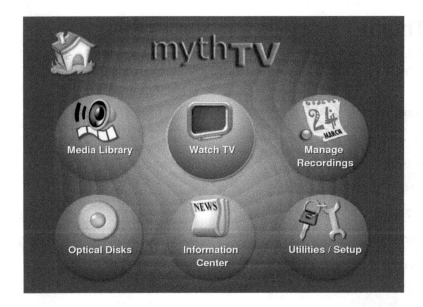

Figure 7-2. *The blue theme*

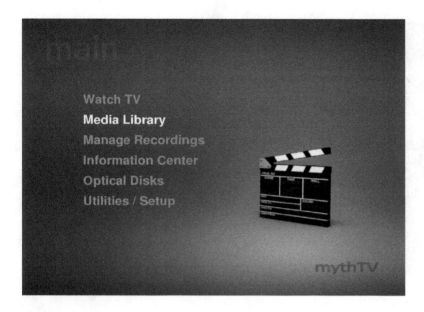

Figure 7-3. *The retro theme*

Figure 7-4. *The minimalist-wide theme*

Figure 7-5. *The iulius theme*

Figure 7-6. *The titivillus theme*

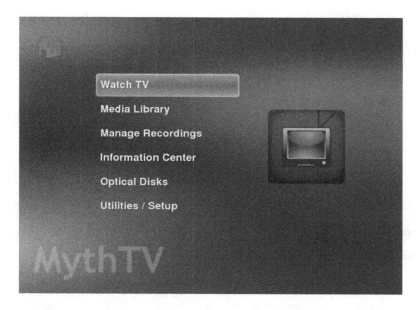

Figure 7-7. *The MythCenter theme*

It turns out that both of us prefer the MythCenter theme, although on poorer-quality televisions, G.A.N.T. can look less glaring. Several settings are independent of the theme yet are strongly related to appearance:

- Qt Style

Note QT is the widget library used by the MythTV frontend. A *widget* is a user interface building block, such as buttons, sliders, progress bars, textboxes, and so forth. You can select which of several sets of widgets you use, which will affect how the user interface looks.

- Font Size

- Language

- Date Format

- Time Format

You can find all of these settings by selecting Utilities / Setup ➤ Setup ➤ Appearance. Those outside of North America will probably want to change the date format to the way they natively write the date. Font sizes can be good to adjust based on the user's quality of vision, television, and distance from the screen. Unfortunately, no preview is provided for these settings, so you might need to experiment.

There are also several types of themes in MythTV; so far, we have talked only about user interface themes—themes that primarily modify how things look, but you can also select on-screen display themes, which modify how the on-screen display during playback appears, and menu themes, which modify where the menu items are.

Finding and Installing Other Themes

You can find an up-to-date listing of some user-contributed themes on the MythTV wiki at
http://www.mythtv.org/wiki/index.php/Themes. In our experience, the highest-quality themes
end up being included in the MythTV distribution, and some are just relatively minor modifications on existing themes. However, reactions to themes vary a lot depending on personal taste.

Installing a theme is rather simple; it's just a matter of downloading it, extracting the
archive, and moving the directory to the appropriate location. When MythTV is installed in
the /usr/local directory (as we detail in this book), the directory where MythTV looks for
themes is /usr/local/share/mythtv/themes/. In this example, we'll show how to download the
blootube theme (pictured in Figure 7-8) from http://www.juski.co.uk/ and install it:

1. Download the theme from http://www.juski.co.uk onto your MythTV box.

2. Extract the archive either using the Archive Manager or using the command line:

   ```
   $ tar xfj blootube.tar.bz2
   ```

3. Move the extracted directory into the MythTV themes directory:

   ```
   $ sudo mv blootube /usr/local/share/mythtv/themes
   ```

4. From the MythTV frontend, choose the blootube theme.

Figure 7-8. *The blootube theme*

Creating Your Own Theme

If none of the available themes is to your exact liking, it's possible to create your own. Many things are customizable through themes in MythTV, including everything from the background image to the structure of the menus. Unfortunately, this means writing a good and complete theme can be a tricky task. There are also other issues to consider: wide-screens, overscan, poorer quality displays, and people with less than perfect vision, to name a few. In the following sections, we'll guide you through some of the common issues you need to consider, and we'll show you a few customizations to an existing well-written theme.

Overscan and Broadcast Television

The first two issues you'll need to consider when working on themes are the aspect ratio of the set you're designing for and the amount of overscan you need to allow for. *Overscan* is the amount of the total picture that isn't visible on a given TV set. Overscan occurred traditionally because of manufacturing tolerances and could amount to as much as 15 to 20 percent of the picture diagonal on older sets. It's much less on newer televisions and can approach zero on LCD and plasma sets. If you're designing a custom theme just for yourself, you can adjust the overscan to suit your particular set. If you want to make a theme that others can use as well, though, you'll have to be careful about how close to the edges your theme places the image.

At Billy Biggs's page about overscan (`http://scanline.ca/overscan/`), he details how overscan works with regular televisions and where the safe parts of the screen are.

4:3 vs. 16:9 and Effectively Testing a Theme

When editing a theme, you'll quickly realize that you want an easy way to test the result of your most recent change. Running a full-screen application (such as the MythTV frontend) along with text and image editors isn't too nice. You will also want to test your theme at the appropriate aspect ratios for TV output, which the standard resolution for your monitor might not be.

The MythTV frontend has an option to run in a window instead of full screen. It also allows you to specify a geometry (like the resolution of your screen) for the window. For 4:3 televisions, the standard resolution on a computer screen is 800×600 pixels, while with 16:9 the resolution is 1280×720 (the same as HDTV 720p). These are the resolutions on which MythTV bases its themes. To run the MythTV frontend in a window at either of these resolutions (so you can see how your theme will look and easily switch between applications, perhaps changing the theme), run one of the following command lines for `mythfrontend`:

```
$ mythfrontend -w -geometry 800x600
$ mythfrontend -w -geometry 1280x720
```

See Figure 7-9 for a MythTV frontend running in a window at 4:3 with the blootube theme, and see Figure 7-10 for a MythTV frontend running as 16:9 with the MythCenter theme.

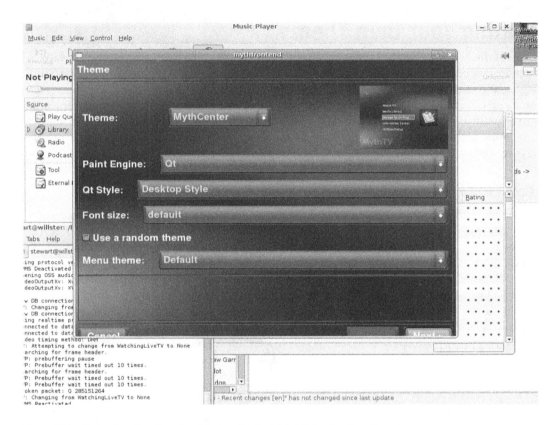

Figure 7-9. *Running the frontend in a window at 4:3 for testing theme appearance*

Figure 7-10. *Running the frontend in a window at 16:9 for testing theme appearance*

Starting a New Theme

When starting a new theme, it's best to base it on an existing good-quality theme and change it as desired. It's also common to want something similar to an existing theme. To copy the MythCenter theme and name it MyTheme, just copy the MythCenter directory. You can do this from the command line like this:

```
$ cd /usr/local/share/mythtv/themes/
$ sudo cp -r MythCenter MyTheme
```

Now, in the MythTV frontend, you can select your theme. To no surprise, it will look exactly like MythCenter. To make it easier to edit the theme without using sudo, you might want to change the file owner like this:

```
$ sudo chown -R mythtv MyTheme
```

Now you can easily edit all the graphics and XML files without using sudo. For user interface themes (such as you're creating here), the details of the theme are in the ui.xml file, and images are stored in other places in the directory (pointed to by options in the ui.xml file). The ui.xml file is an XML file, and standard XML rules apply. Some user interface details are in the qtlook.txt file, which is a key/value-format text file.

Changing the Background Image

One example of a configuration option in the qtlook.txt file is the background image. If you wanted to replace the default background with your own photo (say the one shown in Figure 7-11), you would copy the image into the theme directory (named, for example, ui/background.jpg) and change the corresponding line in the qtlook.txt file from this:

```
"str BackgroundPixmap=ui/background.png"
```

to this:

```
"str BackgroundPixmap=ui/background.jpg"
```

When you start the MythTV frontend with these settings, it looks something like Figure 7-11. You have to be careful in choosing colors for fonts and backgrounds because some screens, such as the program guide, really do need the foreground to be readable to be usable at all (see Figure 7-12). Just because an image looks good with the menus doesn't mean it will look good anywhere else.

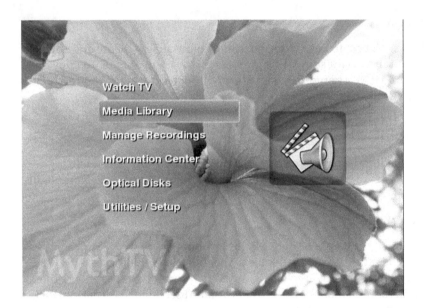

Figure 7-11. *MyTheme—just MythCenter with a different background image*

Figure 7-12. *MyTheme—just MythCenter with a different background image showing the program guide*

It is also important to get the sizes of the images correct. If you look closely at Figures 7-11 and 7-12, you'll notice the background in Figure 7-12 is only the top-left part of what is shown in Figure 7-11. This is because of the background image being full quality and not scaled down to 800×600 pixels. It is quite likely that the resolution being outputted from a video card to a TV is not one of these standard resolutions. It could be raw PAL or NTSC resolution or the native resolution of the display (especially in the case of LCD and plasma screens). Some trial and error might be required to get exactly the desired effect from your theme customizations. We hope these scaling issues will be fixed in a future version of MythTV and that making high resolution themes that scale up and down well will be easier.

Using Basic XML

XML, or eXtensible Markup Language, is a standard for creating "tagging languages" to wrap up data so that programs can easily read and write it. XML supplies the structure, and related standards called document type definitions (DTDs) define which tags take which attributes and which tags can be nested within other tags. Although XML implementations can be quite complex, MythTV's uses of XML are relatively simple, and you can figure out most of what you need to know to modify a theme by looking at the XML. Here's the short set of rules.

A word between angle brackets is a tag (for example <window>). If there are other words, with an equals sign followed by either a number or a quoted string, these are attributes belonging to that tag (for example <window name="programguide">). Each tag has a closing tag (for example <window name="programguide"></window>). Other properties of this can be specified in between the opening and closing tags (for example <window name="programguide"><font name="title" face="Trebuchet MS"></font></window>). A tag can be both an opening tag and a closing tag by having a forward slash before the closing angle bracket (for example <window name= "programguide"/>). A closing tag must be for the last tag that was opened (or the file is invalid).

Editing the ui.xml File

The ui.xml file allows you to specify many specific elements of the user interface. It is mostly divided up into settings for each window. The following is a cut-down ui.xml file specifying a background image for the program guide:

```
<mythuitheme>
  <window name="programguide">
    <container name="background">
      <image name="backup" draworder="0" fleximage="no">
        <context>0</context>
        <position>0,0</position>
        <filename>programbackground.jpg</filename>
      </image>
    </container>
  </window>
</mythuitheme>
```

If an image isn't available (for example, you change the background image for the program guide in MyTheme but forget to copy the image across), the MythTV frontend will print a warning to the terminal where you started the frontend. This can be useful in debugging your theme changes, and as you might expect, it implies that you should be starting your frontend manually from a terminal window while you're debugging—if you let the system start it from a boot-time script, as you would in regular operation, those messages might be lost or hidden in a log file somewhere.

Editing the theme.xml File

The theme.xml file allows you to specify some global characteristics of the theme. It's possible to define the specifics of fonts here instead of in each window in the ui.xml file. This saves some processing time for MythTV when setting up which fonts it needs to draw a particular screen.

Changing Fonts

You specify fonts either in the theme.xml file or in the ui.xml file. Often in the ui.xml file, the font will be for a specific UI element.

Changing Buttons

The button images dictate how unselected and selected UI elements appear on the screen. The easiest way to manipulate these is to change the images. Depending on the degree of modification you want to make, you might also need to edit the XML description in theme.xml to account for the different size. Figure 7-13 shows an incredibly ugly change to the default button image. Figure 7-14 shows the area for the text in the genericbutton part of theme.xml having been changed to center the text in the new button image.

Figure 7-13. *MyTheme with a different image for the button*

Figure 7-14. *MyTheme with a different image for the button and the text area changed*

Changing Title and Watermark Images

The final customization we'll cover is title and watermark images. The title image appears on the top left of the menu in MythCenter. You can change the location of this in theme.xml (using x,y coordinates). The watermark is the image displayed (on the right for MythCenter) in the frontend to show what option you currently have selected. Again, it's a simple edit to change the image files. In Figure 7-15, we've changed the title image for the main menu to a photo of a leaf and the watermark for Watch TV to a photo of a flower. For your own theme, you would likely choose more appropriate images. Using images in the PNG format has the advantage of being able to have an alpha channel that blends the background with the foreground, like the original images in the MythCenter theme do. You can especially notice the lack of an alpha channel in the leaf image in the way it covers up the left of the Watch TV button.

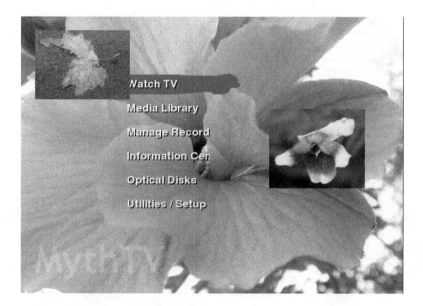

Figure 7-15. *MyTheme with new title and watermark images*

Using Tools for Creating and Editing Images

Several free, open source tools come with Ubuntu Linux that are good for editing and creating images. Often these tools are available for other platforms too. Gimp (http://www.gimp.org/) is an image manipulation program often compared to Adobe Photoshop. Inkscape (http://www.inkscape.org) is a vector graphics editor often compared to Adobe Illustrator, Adobe FreeHand, and CorelDRAW. Many artists are now using these tools full-time to produce the graphics and images that come with Ubuntu. For producing 3D graphics, many people use the Blender program (http://www.blender.org). Do be aware, though, that it can take some time to get used to these powerful pieces of software and that producing quality, original graphics isn't the easiest task in the world.

Conclusion

In this chapter, we covered selecting a theme for your MythTV frontend; we also showed some simple steps to help you create your own. We didn't go into on-screen display or menu themes, although the basic ideas are the same and largely depend on producing correct XML for the MythTV frontend to interpret. The MythTV theme system is designed to give theme builders a lot of flexibility to change the interface. However, some changes just aren't possible in the current infrastructure (although these are quite rare for the average theme developer). If you have suggestions for improvements, people are probably willing to listen! For more information, always check out the mailing lists and wikis.

■ ■ ■

## CHAPTER 8

Running Remote Frontends

This chapter discusses how to build remote frontends for your MythTV installation. Up until now, we have shown how to run the MythTV frontend on the same machine as the backend, but you don't have to set up your MythTV installation that way. In fact, you can have more than one frontend if you want. You might want a remote frontend for a variety of reasons. Perhaps your backend machine is too big or too loud to be in your living room. Perhaps you want to be playing more than one video at once in different parts of the house.

This chapter will cover building three types of remote frontends. First we'll talk briefly about Linux-based frontends, then we'll talk about Xbox-based frontends (which are a slightly more specialized version of the generic Linux frontend), and finally we'll cover Macintosh frontends. We'll also briefly cover options for viewing videos on machines other than your backend without going the whole way and installing an entire frontend.

Getting Ready for Other Frontends

You need to take some preparatory steps before you can run frontends on machines other than the machine on which the backend runs. First, you need to configure the network to have the MythTV backend machine on the same static IP address all the time. Second, you need to configure the MythTV backend to be accessible to the network. Finally, you need to tweak the MySQL configuration to work on the network as well. If you change the MythTV network configuration and find that it's still not working, check your firewall settings.

Network Configuration

Obviously, you need a network connection between your MythTV backend machine and the various frontend machines on your network. You probably already have your backend machine on the network so you can get TV guide data information from the Internet, so this shouldn't need a lot of work. However, it is also important that you have a static IP address for your backend machine.

It's hard to describe how to configure your network connection for the backend machine to have a static IP address in a book like this. You can use a variety of command-line tools, or you can use graphical tools, which of course vary depending on what window manager you are using. Your best bet it to check out *Beginning Ubuntu Linux: From Novice to Professional, Second Edition* (Apress, 2007) or to search online for descriptions of how to set the static IP for your network card. You should remember to ensure that you use an IP address that your DHCP server (on small home LANs, this is usually on your router) isn't handing out, though, because otherwise you will end up with two machines fighting over that address.

Backend Configuration

You first need to configure the backend software to use an IP address that is accessible from the network (it defaults to 127.0.0.1 or localhost, which is not network accessible). You do this by running mythtv-setup and going to the general preferences. The first page of that wizard looks like Figure 8-1.

Tip You might find that you don't have a mythtv-setup program. On some distributions, it is named mythtvsetup.

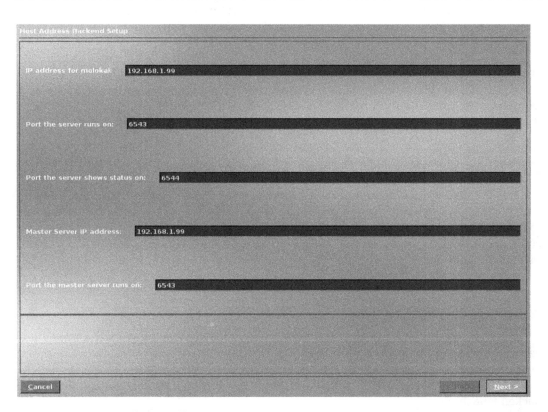

Figure 8-1. *Setting the IP address of the backend*

You can see here that the IP address of this backend is 192.168.1.99. Yours has probably defaulted to 127.0.0.1 and should be changed to the IP address of the machine running the backend. You can find out the IP address of the machine that is running the backend by going to a command prompt and running the ifconfig command. In this example, we know that the name of the interface is eth0; if you don't know the name of the interface, then just omit that portion of the command line:

```
$ ifconfig eth0
eth0      Link encap:Ethernet  HWaddr 00:13:20:9D:71:A5
          inet addr:192.168.1.99  Bcast:192.168.1.255  Mask:255.255.255.0
          inet6 addr: fe80::213:20ff:fe9d:71a5/64 Scope:Link
          UP BROADCAST RUNNING MULTICAST  MTU:1500  Metric:1
          RX packets:360121 errors:0 dropped:0 overruns:0 frame:0
          TX packets:644094 errors:0 dropped:0 overruns:0 carrier:0
          collisions:0 txqueuelen:1000
          RX bytes:25279027 (24.1 MiB)  TX bytes:935139149 (891.8 MiB)
```

MySQL Configuration

Now, you need to configure MySQL to be accessible on your network. The first step is to tell MySQL to listen on the IP address that is accessible to the network, in the same way you did for the MythTV backend. However, MySQL doesn't have a pretty GUI configuration tool, so you'll need to edit the /etc/mysql/my.cnf file yourself. Run your favorite editor to do this, but remember to use sudo so that you have root permissions for your edits.

Look for a stanza like this in the configuration file:

```
# Instead of skip-networking the default is now to listen only on
# localhost, which is more compatible and is not less secure.
bind-address            = 192.168.1.99
```

Here you can see that we set the bind address to the same IP address that we used for the MythTV backend in the previous section. You should edit this to be whatever the IP address for your backend is.

Next, you need to give the new frontend permissions to access the MySQL server. You'll find a sample script to do that in the MythTV source code in a directory called database. Here is a modified version that does only the bare minimum to get a new frontend working. We've highlighted the parts you'll need to tweak for your setup in bold.

```
GRANT ALL ON mythconverg.* TO mythtv@remoteip IDENTIFIED BY "newpassword";
FLUSH PRIVILEGES;
GRANT CREATE TEMPORARY TABLES ON mythconverg.* TO mythtv@remoteip ➥
IDENTIFIED BY "mythtv";
FLUSH PRIVILEGES;
```

Here you need to replace remoteip with the IP address of the machine the new frontend is going to run on or with '%' if you either don't care what IP is connecting to MySQL from or are using DHCP. If you're installing a machine specifically to run your new frontend, then you might need to return to this section later when you know the IP address you have allocated to the machine. You also need to specify a remote password (newpassword), which the remote machine will use to connect. It's important to note that the password does not have to be the same for each IP, even if the username is the same, which is very cool.

You can get a dump of the users who are allowed to connect to the MySQL database by asking MySQL like this:

```
$ mysql -u root -p mysql
Enter password: <your superuser password for MySQL>
Reading table information for completion of table and column names
You can turn off this feature to get a quicker startup with -A

Welcome to the MySQL monitor.  Commands end with ; or \g.
Your MySQL connection id is 124 to server version: 5.0.19-Debian_3-log

Type 'help;' or '\h' for help. Type '\c' to clear the buffer.

mysql> select * from user \G;
*************************** 1. row ***************************
...
*************************** 2. row ***************************
                 Host: 192.168.1.100
                 User: mythtv
             Password: 0476fc996afffe24
          Select_priv: N
          Insert_priv: N
          Update_priv: N
          Delete_priv: N
          Create_priv: N
            Drop_priv: N
          Reload_priv: N
        Shutdown_priv: N
         Process_priv: N
            File_priv: N
           Grant_priv: N
      References_priv: N
           Index_priv: N
           Alter_priv: N
         Show_db_priv: N
           Super_priv: N
Create_tmp_table_priv: N
     Lock_tables_priv: N
         Execute_priv: N
      Repl_slave_priv: N
     Repl_client_priv: N
             ssl_type:
           ssl_cipher:
          x509_issuer:
         x509_subject:
        max_questions: 0
          max_updates: 0
      max_connections: 0
...
```

We've used the \G option with the SELECT statement, which presents the information in this list format instead of the traditional ASCII art table form, because the data is too wide to fit on a book page or screen. You can see here the details for one of our remote frontends, which has an IP address of 192.168.1.100.

Specific Types of Frontend

Now we can discuss how to set up specific types of frontends. Each of the following discussions is separate, and you can skip the sections you are not interested in without running the risk of ending up with a setup that doesn't work.

Running Another Linux Frontend

You build Linux frontends in the same manner as you built the MythTV machine in Chapters 2 and 3. When you're compiling the MythTV code, you don't need to build the backend as well as the frontend, and the backend on these machines won't be used. Unfortunately, there is no easy way to disable compiling the backend code, so we'll just show how to compile it and then never use the backend.

Frontends don't need their own copies of MySQL running or XMLTV and the guide data setup, so don't install them on the new frontend machine. Here is a brief overview of the relevant steps to build MythTV for your new frontend. You can find a more complete discussion in Chapters 2 and 3.

```
$ sudo apt-get build-dep mythtv
Password:
Reading package lists... Done
Building dependency tree... Done
The following NEW packages will be installed:
...
Need to get 37.9MB of archives.
After unpacking 135MB of additional disk space will be used.
Do you want to continue [Y/n]? y
Get:1 http://security.ubuntu.com dapper-security/main ➡
libfreetype6-dev 2.1.10-1ubuntu2.2 [677kB]
...
$ sudo apt-get install libqt3-mt-mysql
Reading package lists... Done
Building dependency tree... Done
The following NEW packages will be installed:
  libqt3-mt-mysql
0 upgraded, 1 newly installed, 0 to remove and 0 not upgraded.
Need to get 51.9kB of archives.
After unpacking 139kB of additional disk space will be used.
Get:1 http://archive.ubuntu.com dapper-updates/universe ➡
libqt3-mt-mysql 3:3.3.6-1ubuntu6 [51.9kB]
Fetched 51.9kB in 1s (47.0kB/s)
Selecting previously deselected package libqt3-mt-mysql.
```

```
(Reading database ... 82299 files and directories currently installed.)
Unpacking libqt3-mt-mysql (from .../libqt3-mt-mysql_3%3a3.3.6-1ubuntu6_i386.deb) ...
Setting up libqt3-mt-mysql (3.3.6-1ubuntu6) ...
```

```
$ wget "http://www.mythtv.org/modules.php?name=Downloads&d_op=getit&lid=129"
--13:25:10--  http://www.mythtv.org/modules.php?name=Downloads&d_op=getit&lid=129
          => 'modules.php?name=Downloads&d_op=getit&lid=129'
Resolving www.mythtv.org... 140.211.167.131
Connecting to www.mythtv.org|140.211.167.131|:80... connected.
HTTP request sent, awaiting response... 302 Found
Location: http://ftp.osuosl.org/pub/mythtv/mythtv-0.20.tar.bz2 [following]
--13:25:25--  http://ftp.osuosl.org/pub/mythtv/mythtv-0.20.tar.bz2
          => 'mythtv-0.20.tar.bz2'
Resolving ftp.osuosl.org... 64.50.236.52, 64.50.238.52
Connecting to ftp.osuosl.org|64.50.236.52|:80... connected.
HTTP request sent, awaiting response... 200 OK
Length: 12,380,677 (12M) [application/x-tar]

100%[===============================>] 12,380,677    158.27K/s    ETA 00:00

13:44:36 (158.41 KB/s) - 'mythtv-0.20.tar.bz2' saved [12380677/12380677]
```

Notice that this wget download redirects to a file named mythtv-0.20.tar.bz2 and that it will likely be slightly different for newer releases. Now extract the download, and then compile:

```
$ tar --bzip -xf mythtv-0.20.tar.bz2
$ cd mythtv-0.20
$ ./configure
...
$ make
...
$ make install
...
```

After doing all this, you have all the software you need to run a frontend, although you still need to configure the remote control using lirc, as discussed in Chapter 3. You might need to add /usr/local/lib/ to /etc/ld.so.conf if you see errors about libraries not being found. This is as simple as editing this file as root and adding the new line. After that, you need to rebuild the library cache with this command:

```
$ sudo ldconfig
```

Now that you have all the software ready, you need to run mythfrontend to start configuring the frontend. You'll see a selection screen that asks you to pick your language (see Figure 8-2).

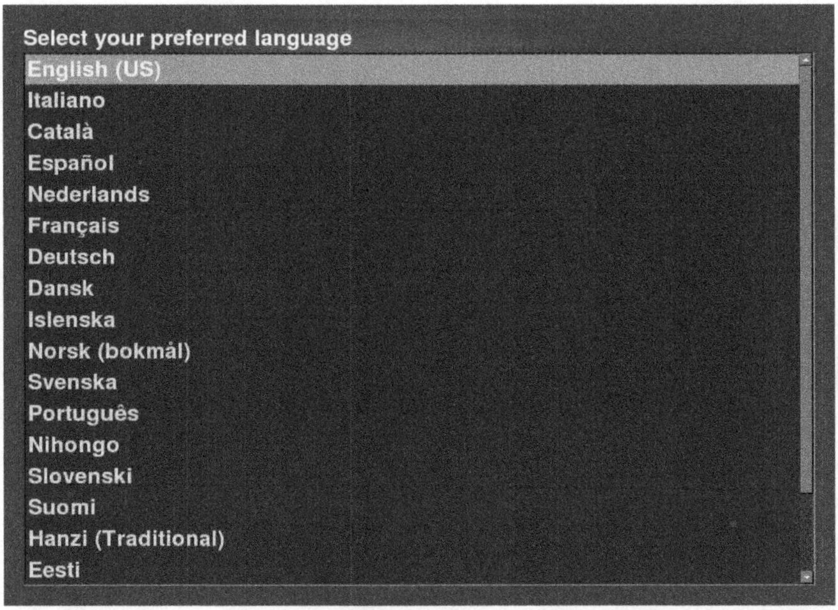

Figure 8-2. *mythtv-setup prompting for your language choice*

Next you're prompted for database configuration (see Figure 8-3).

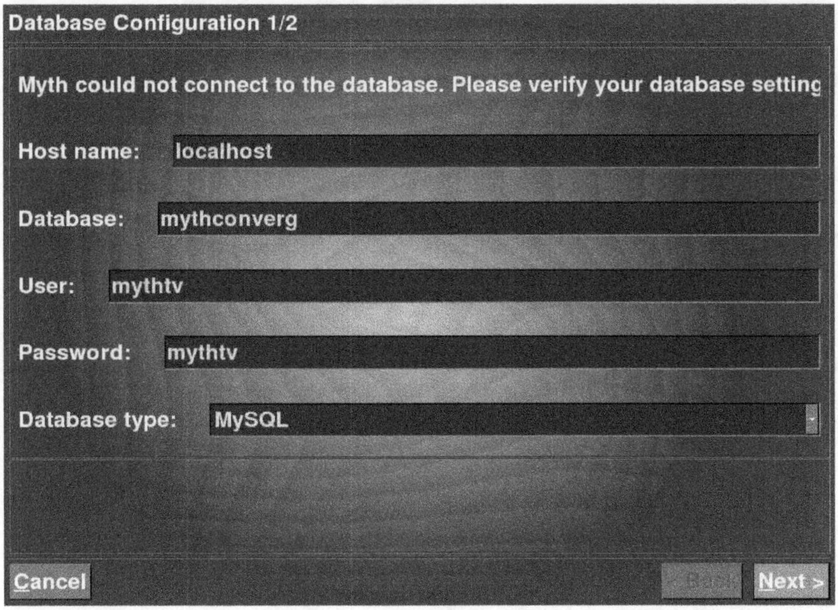

Figure 8-3. *Database configuration defaults*

You need to change these default values to whatever is correct for your network (see Figure 8-4).

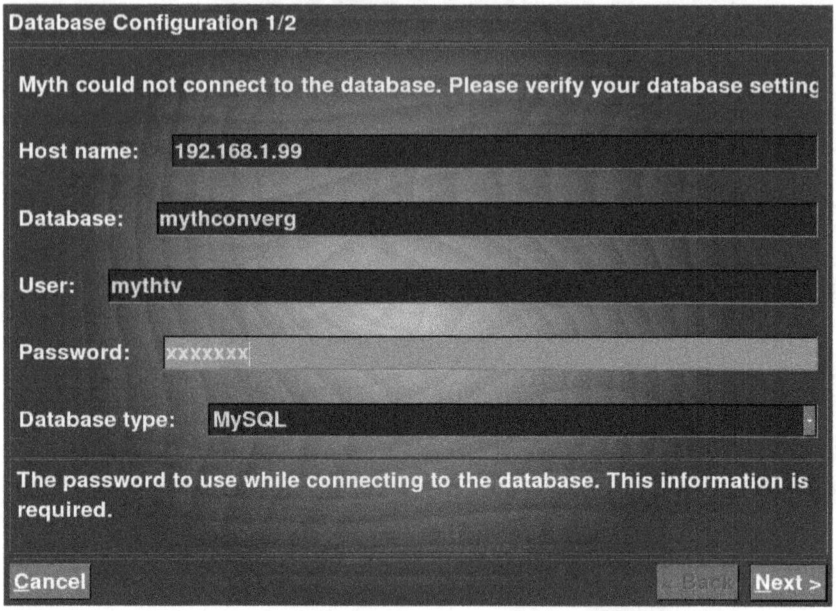

Database Configuration 1/2

Myth could not connect to the database. Please verify your database setting

Host name: 192.168.1.99

Database: mythconverg

User: mythtv

Password: xxxxxxx

Database type: MySQL

The password to use while connecting to the database. This information is required.

Cancel   Back   Next >

Figure 8-4. *Setting the IP address of the backend, as well as the username and password*

Here you need to set up the IP address (or DNS name) of the machine running the backend, the name of the MySQL user who is going to connect from this frontend, and the MySQL password that is valid for that username, connecting from the IP address that the frontend is configured with. (As always, remember that MySQL usernames are not connected to Linux usernames; don't get confused between them.) You can set some other configuration options that appear on the next screen, as shown Figure 8-5.

The most interesting piece here is the custom identifier for the frontend. If the frontend is using DHCP to get its network configuration, then it's possible that the IP address of the frontend is going to change. MythTV by default saves the configuration settings for a frontend based on its IP, but this doesn't work if the IP address of the machine is going to change. In that case, you're better off setting a text description of the machine, as we have done here.

Database Configuration 2/2

☑ Use custom identifier for frontend preferences

Custom identifier: bedroom

☐ Use Wake-On-LAN to wake database

An identifier to use while saving the settings for this frontend.

Cancel   < Back   Finish

Figure 8-5. *Further configuration for the database*

When you're finished, then you should be able to run the frontend using mythfrontend. The theme won't have been configured yet (configuring themes is covered in the previous chapter of this book), so you'll end up with the default, which is shown in Figure 8-6.

Figure 8-6. *The new frontend running with the default theme*

Now, your new Linux frontend should be ready to go. We won't discuss here how to get DVD playback working for full-sized Linux frontend machines (although we do discuss it for Xbox Linux frontends in the next section), because that is the focus of Chapter 12.

Running a Frontend on an Xbox

We built a Microsoft Xbox frontend for a MythTV setup for a few reasons: we already had an Xbox lying around unused; the Xbox hardware is very cheap; there is plenty of remote control hardware available for the Xbox that is simple to install; and an Xbox can natively display on a regular TV set without any additional hardware, which worked well for the room to which we wanted to add a frontend. In this chapter, we're referring to Microsoft's original Xbox hardware, not the newer Xbox 360, which has not yet been hacked to run Linux. We used a Microsoft DVD remote control set for the remote control on this frontend. Building an Xbox-based frontend is really a special case of building any other Linux frontend.

The first step is to install a Linux distribution on your Xbox. This isn't as hard as it sounds, because the Xbox Linux project at `http://www.xbox-linux.org/wiki/Main_Page` has already solved this problem. We chose Xebian, because it's close to the Debian and Ubuntu distributions, which we use on our other systems. You can find Xebian at `http://www.xbox-linux.org/wiki/Xebian`. For Xebian to run on our version of Xbox, we needed to install a Cromwell BIOS, which means that the Xbox can no longer play Xbox games. That's not such a big deal given that we don't play computer games. There are other options here, and if you want your Xbox to remain as a game-playing machine as well, then you should read up on your options at the Xbox Linux site.

If you don't want to go the Xebian route, then you might be interested in the Xbox Media Center (XBMC) project, which also allows you to have an Xbox-based MythTV frontend. You can find more details of this at `http://sourceforge.net/projects/xbmcmythtv/`, so we won't discuss it anymore here.

We won't provide a complete Xbox Linux tutorial here, because there are plenty of those on the Internet already, including the Xbox Linux project and Xebian sites that we have mentioned already. We recommend that you walk through the Xbox Linux installation details on the Xbox Linux site at `http://www.xbox-linux.org`. They can even direct you to volunteers who will help you with the Linux install if you need that.

One note before we get started discussing the Xbox installation. We built an Xbox frontend because we already had a modified Xbox lying around. If we hadn't owned the hardware, then we would have seriously considered using an old laptop or other machine instead. The main reason for our hesitation to rush out and purchase an Xbox is that the machines are quite loud when running, and that isn't great in a room where you want to watch TV. On the other hand, the Xbox hardware is cheap and does work nicely with standard-definition televisions.

The details on how to install Xebian are at `http://www.xbox-linux.org/wiki/Xebian_HOWTO#Finding_and_downloading_the_latest_version` and are fairly easy to follow. Remember, when you configure your Xbox frontend, you need to set up the machine with either DHCP or an IP address that is going to work on your network. You can find a more detailed discussion about this in the "Network Configuration" section.

Note The DVD player in the Xbox is designed to play only commercially produced DVDs. This has the rather nasty side effect that not all burned CD-ROMs or DVDs work with the Xbox DVD drive. This can cause you problems when you try to install Xebian or any other Xbox Linux distribution. See the Xebian description of the problem at `http://www.xbox-linux.org/wiki/Xebian_HOWTO#Media_matters` for more details.

Once you have Xebian installed, which took us about 30 minutes not including burning the CD-ROM and modifying the Xbox hardware, you are ready to follow the setup instructions for MythTV's frontend. Because of the slow CPU on the Xbox, you're best off installing from packages instead of building the whole thing from source yourself (source is an option if you really want, though). To install the MythTV packages, just follow the instructions on the Xbox Linux site at `http://www.xbox-linux.org/wiki/MythTV_on_Xebian_HOWTO#Install_MythTV`. To briefly summarize those instructions, do something like the following:

```
$ apt-get update
$ apt-get install mythtv-frontend
```

This will leave you with a `mythfrontend` that should just work, including remote control support for the Microsoft Xbox DVD playback–style remote controls. If you don't have a Microsoft Xbox DVD–compatible remote control, then you might need to modify the `lirc` configuration for the system in order to get the remote control to work. In that case, your best bet is to do a Google search for an existing `lirc` configuration file before creating your own.

You'll need to configure the database on the backend machine to allow the Xbox to connect to the database before anything will work. We discussed this in the "MySQL Configuration" section earlier in this chapter, along with the need to change your MySQL and MythTV backend configurations slightly from the defaults.

Once the install is finished and a database account is created, then you'll be able to run `mythfrontend` as the user "live" and see whether it works. ("live" is the name of the default account on a Xebian system.)

We found we had trouble with error messages about old MySQL clients:

```
xebian:~# mysql -h backendip -u mythtv -p mythconverg
Enter password:
ERROR 1251: Client does not support authentication protocol requested by
  server; consider upgrading MySQL client
```

In this example, `backendip` is either the DNS name of the machine running the MythTV backend or the IP address of that machine. It's quite possible by the time you read this, though, that the Xebian-packaged version of the MySQL client will have been upgraded and that you won't see this message at all. If you experience this symptom, then you can fix the problem by using an old-style password for that one client by executing this SQL on your MySQL server:

```
update user set password = old_password('newpassword') where host = 'remoteip';
```

where `remoteip` is the IP address of that client machine, `newpassword` is the password the client should present to the server, and `old_password` is the name of a function provided by MySQL, which converts it appropriately. You can find more details of the problem and the other possible solutions at the MySQL website at `http://dev.mysql.com/doc/refman/4.1/en/old-client.html`.

You can test that the frontend works by starting it up and seeing whether there are any recordings in Media Library ➤ Watch Recordings. Finally, after you've done all this, you just need to get DVD playback working. To do this, you need to install the DVD plug-in. Install some other plug-ins while you're there:

```
$ apt-get install mythdvd mythgallery mythgame mythmusic mythvideo
```

This will add DVD playback support, a picture display application, arcade game simulation, MP3 playback, and playback of other video files (such as MPEG files and AVIs). Note that many of these modules expect to get their content from the file system. For example, the MythMusic plug-in expects to find the MP3s to play somewhere on the local file system. This means that if you want to have that sort of content stored centrally (perhaps on the same machine as the MythTV backend), then you'll need to export those file systems that contain content using a mechanism such as NFS—MythTV frontends retrieve their recorded programs via a built-in networking protocol, but plug-ins need to make their own arrangements (or, more properly, you do). You can read more about these plug-ins in Chapter 9.

You also need to make sure the /dev/dvd device file exists. If it doesn't, then just make a symbolic link to the /dev/cdrom file like this:

```
$ ln -s /dev/cdrom /dev/dvd
```

DVD playback actually works better than the experiences we had originally with a full-blown PC. The main reasons for this are that the remote control configuration as shipped by the MythDVD package and Xebian is much better tailored, volume control is not something we do with our Xbox (it's done by external hardware), and there is a physical eject switch on the Xbox's DVD drive that always ejects the disc, regardless of what the Xbox is currently doing.

Tip Remember that in order to play region-encoded DVDs or encrypted DVDs, you need to install the libdvdcss library separately from xine itself. They are packaged separately because of possible legal implications with libdvdcss. Please make sure it's legal in your country first. If it is legal, then installation is as easy as this:

```
$ sudo apt-get install libdvdcss2 libdvdcss2-dev
```

Finally, the history of how the Microsoft Xbox was hacked to run Linux is a story we find quite interesting; you can find an excellent introduction to Microsoft's Xbox hardware and how Linux came to be run on it in *Hacking the Xbox: An Introduction to Reverse Engineering* by Andrew Huang (O'Reilly, 2003).

Building a Macintosh Frontend

Setting up a Macintosh frontend is much like the other frontends we have discussed. The first step is to download a precompiled version of `mythfrontend`. We got ours from `http://collectivity.goof.com/`; specifically, we used `http://collectivity.goof.com/articles/2006/09/27/mythfrontend-0-20-for-all-mac-os-x-variants`. Download that file onto your Macintosh, and then double-click it to extract it (if you used Safari, then you can do that in the Downloads window). You can see an example in Figure 8-7. You'll end up with a disk image that Finder will mount for you. This can take quite a while, because the disk image is quite large.

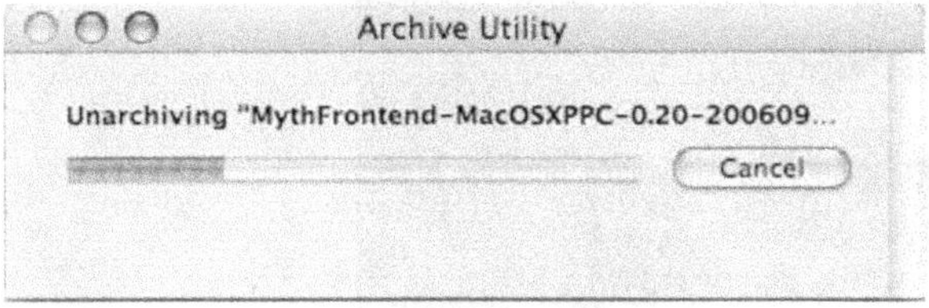

Figure 8-7. *Extracting the disk image*

Once you've extracted the disk image, you can copy the `mythfrontend` application that is inside it to wherever you want on your machine. We put it in the `Applications` folder and then created a shortcut on the Dock for it. When you start the application, you'll get the same configuration questions that you do for the other frontends. First pick a language (see Figure 8-8).

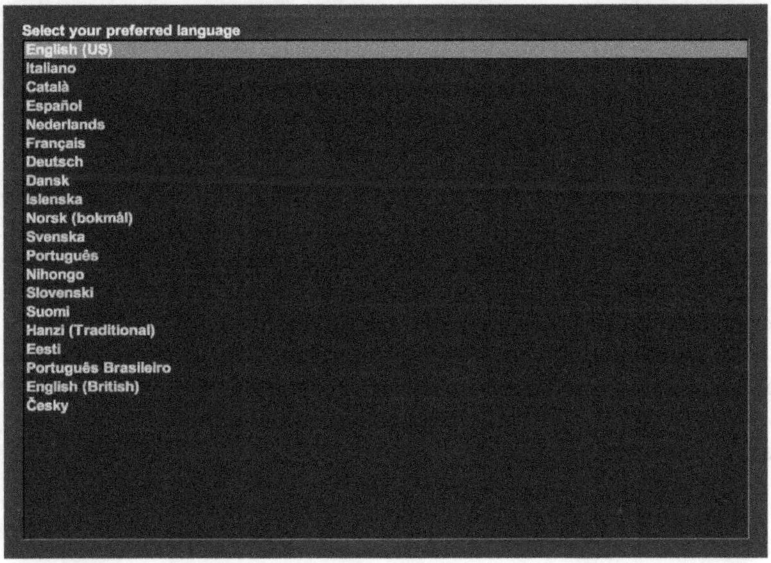

Figure 8-8. *Selecting a language*

Then, enter the configuration information for your backend (see Figure 8-9).

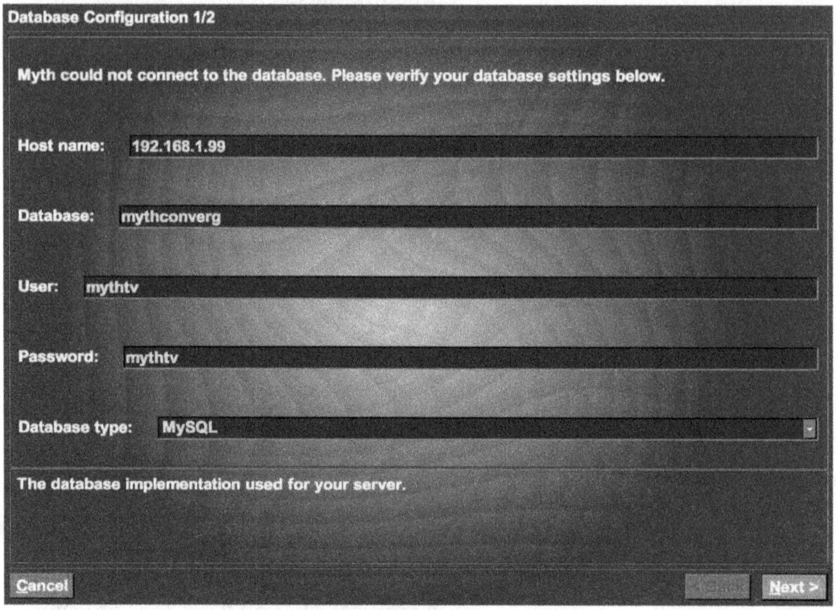

Figure 8-9. *Configuring the backend details*

Here you need to set up the IP address (or DNS name) of the machine running the backend, the name of the MySQL user who is going to connect from this frontend, and the MySQL password that is valid for that username, connecting from the IP address that the frontend is configured with. You can set some other configuration options that appear on the next screen, as shown in Figure 8-10.

The most interesting piece here is the custom identifier for the frontend. If the frontend is using DHCP to get its network configuration, then it's possible that the IP address of the frontend is going to change. MythTV by default saves the configuration settings for a frontend based on its IP, but this doesn't work if the IP address of the machine is going to change. In that case, you're better off setting a text description of the machine, as we have done here.

When you're finished, you'll end up in the MythTV user interface, with the default theme. (Chapter 7 covered themes if you want more details on those.) Depending on the version of the mythfrontend you downloaded, you should also find that the Apple remote control just works without needing extra configuration.

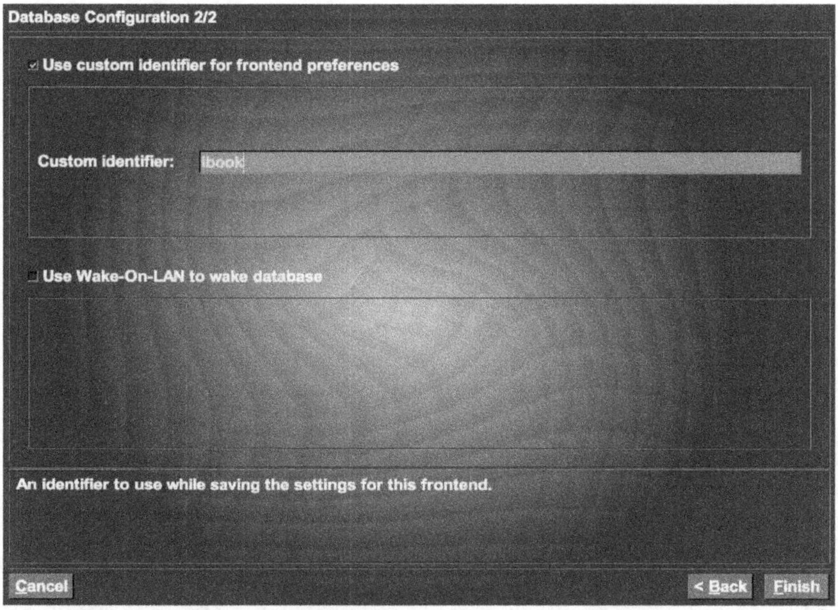

Figure 8-10. *Further configuration for the database*

Windows

There is a Windows version of `mythfrontend` as well, with its home page located at `http://winmyth.sourceforge.net/`. However, it is early in its development and not up to the standard of the other `mythfrontend` options at the moment. We'll therefore not cover how to install it here and will instead recommend that you pursue some of the other viewing options, discussed next, if you want to watch your recordings on Windows machines.

Other Remote Viewing Options

You have some options for viewing recordings from MythTV remotely that don't include setting up a frontend, and the following sections are devoted to these options. These options should be attractive to you if setting up a frontend is too hard (for example you want to watch the videos on a Windows machine) or if there is a low-bandwidth connection between the backend and the machine you want to watch the videos on (for example you want to watch videos over the Internet).

Transcode and Download

Transcoding your MythTV recordings will have two effects: it will make the files smaller (which can be a big benefit if your network link is slow), and it converts the video to formats that are better supported by generic video players such as those you probably already have installed. Chapter 5 covered trancoding videos, so we won't cover that again here. It is worth saying that the divx format that `nuvexport` can export to is probably the best supported video format for various operating systems, so it is a good first choice when you are experimenting with what your playback machine supports. You can also set up `nuvexport` to transcode videos automatically, which was also covered in Chapter 5.

If you would like to make the transcoded videos available on the Internet, then the easiest way is to have Apache serve the contents of the directory that your transcoded files are saved to and then access them remotely via HTTP. If you want to make the videos available on your LAN only, then you might consider NFS or SMB depending on your operating system choice.

Use a Player That Understands MythTV Formats

The drawback to transcoding video is that it is expensive in CPU time, which means you might find you have a hard time keeping up with the content you are recording if you are recording a lot. In that case, you might just want to copy the files in the MythTV recording format and then find a player that supports playing back the MythTV video format. The catch with this, of course, is that the files will be bigger.

One problem with this is that the MythTV NUV format is actually a container format that can include video compressed with a number of different codecs. MPlayer is a good choice of cross-platform video player that supports NUV. You can download copies of MPlayer for Linux, Windows, and Macintosh at `http://www.mplayerhq.hu/design7/dload.html`. Installing MPlayer is well documented at `http://www.mplayerhq.hu/DOCS/HTML/en/install.html`.

Another catch is that the filenames for the MythTV recordings are hard to understand, because they are not based on the show titles. For example, here is a subset of the listings in one of our video directories at the moment:

```
$ ls | head -10
1002_20060730203000.mpg
1002_20060730203000.mpg.png
1002_20060806203000.mpg
1002_20060806203000.mpg.png
1005_20060729180000.mpg
1005_20060729180000.mpg.png
1005_20060730180000.mpg
1005_20060730180000.mpg.png
1005_20060801173000.mpg
1005_20060801173000.mpg.png
```

The `.png` files here are thumbnails to be displayed in various user interfaces, and the `.mpg` files are MPEG-2 video from one of our Hauppauge PVR 350 cards (which records natively in MPEG-2). These filenames can make it hard to determine which video file to play with MPlayer. There are three options to determine the filenames that map to the shows you want to watch. We'll cover them in the following sections.

Using nuvexport to Determine Filenames

`nuvexport` can tell you the filename of videos. Chapter 5 covered `nuvexport` much more completely (including how to install it and the most useful command-line options), but here is a simple example of how to determine the filename of a video based on a search:

```
$ nuvexport --title "Whose Line Is It Anyway?" --search-only
Loading MythTV recording info.
99%

Matching Shows:

    title:  Whose Line Is It Anyway?
 subtitle:  Untitled
   chanid:  1052
   starts:  20060724223000
     ends:  20060724230000
 filename:  /var/video//var/video/1052_20060724223000.mpg

    title:  Whose Line Is It Anyway?
 subtitle:  Untitled
   chanid:  1052
   starts:  20060725223000
     ends:  20060725230000
 filename:  /var/video//var/video/1052_20060725223000.mpg
```

The --search-only flag tells nuvexport to not transcode the video and just output search results instead. You can see here that we have two episodes of *Whose Line Is It Anyway?* recorded, and their filenames are in the output next to the filename label.

Using MythWeb to Determine Filenames

MythWeb, which we discuss in Chapter 11, gives you a graphical way of finding recorded shows, and if you hover over the thumbnail of the video, then the filename of the recording is included in the URL. In fact, clicking the thumbnail will download the video and either save it on disk or start playing it in a video player depending on your browser preferences. See Chapter 11 for more details of MythWeb.

Using User Jobs to Determine Filenames

You can also create a MythTV user job that creates a symlink from a descriptive name for a show to the MythTV recording file. Chapter 5 includes complete coverage of user jobs, but, briefly, a user job like the following one will create a symlink between the real video file and a file with a descriptive name in /data/links/:

```
$ ln -s %DIR%/%FILE% /data/links/`echo %TITLE%-%SUBTITLE% | sed 's/ \//_/g'`
```

The sed portion of this command line is needed to turn spaces and other illegal filename characters in the destination filename into underscores.

MythWeb

Finally, for these other options, MythWeb, which is discussed in Chapter 11, offers a download link for videos. Click the thumbnail next to a video, and your browser will either download or start playing the video, depending on your browser settings.

Conclusion

In this chapter, we covered how to set up frontends on machines other than the one running your MythTV backend. We also talked about the generic setup steps and then how to install frontends on Linux, on Xbox machines running Linux, and on Macintosh machines. Finally, we talked about other options if you don't want to go the whole way to installing a frontend.

The next chapter covers all the plug-in functionality available in MythTV, which lets you perform tasks such as check the weather, manage and play your MP3s and videos, and do cool stuff like look after your Netflix queue. Read on for details about that and much more.

## CHAPTER 9

Installing Other Plug-Ins

MythTV has an extensible architecture that allows for plug-ins to be written to provide functionality not implemented by the core team of MythTV developers. Also, an official download contains a good selection of plug-ins. In addition to the ones provided on the MythTV site, quite a few third-party plug-ins are available since the interface to write plug-ins is public.

This chapter will discuss the majority of the plug-ins provided by the MythTV development team, although some of them are reserved for separate discussion in other chapters of this book. When we come across a plug-in discussed elsewhere, we will refer you to the right chapter so you can learn more.

Installing MythTV Plug-Ins

To get MythTV's official plug-ins to compile, you need to install some dependencies. We have already discussed this in Chapter 3, but in case you skipped over it there, here are the relevant steps:

```
$ sudo apt-get builddep mythplugins
```

To install the MythTV provided plug-ins, download the MythTV plug-ins `tar` archive from the MythTV site downloads area at http://www.mythtv.org/modules.php?name=Downloads&d_op=getit&lid=130 (130 is the proper `lid` parameter for version 0.20; this will change in later releases). You then extract the downloaded file and compile it like this:

```
$ wget "http://www.mythtv.org/modules.php?name=Downloads&d_op=getit&lid=130"
...
HTTP request sent, awaiting response... 302 Found
Location: http://ftp.osuosl.org/pub/mythtv/mythplugins-0.20a.tar.bz2 [following]
...
21:16:11 (154.52 KB/s) - `mythplugins-0.20a.tar.bz2'saved [17852673/17852673]
$ tar --bzip -xf mythplugins-0.20a.tar.bz2
$ cd mythplugins-0.20a
$ ./configure --enable-transcode --enable-vcd --enable-aac --enable-festival

Configuration settings:
```

```
         MythBrowser    plugin will be built
         MythControls   plugin will be built
         MythFlix       plugin will be built
         MythDVD        plugin will be built
         MythGallery    plugin will be built
         MythGame       plugin will be built
         MythMusic      plugin will be built
         MythNews       plugin will be built
         MythPhone      plugin will be built
         MythVideo      plugin will be built
         MythWeather    plugin will be built
         VCD            support will be included in MythDVD
         Transcode      support will be included in MythDVD
         OpenGL         support will be included in MythGallery
         EXIF           support will be included in MythGallery
         OpenGL         support will be included in MythMusic
         FFTW v.3       support will be included in MythMusic
         SDL            support will be included in MythMusic
         AAC            support will be included in MythMusic
         FESTIVAL       support will be included in MythPhone
$ qmake mythplugins.pro
$ make
...
$ sudo make install
...
```

Note You might notice here that the plug-ins download is called `mythplugins-0.20a`, not just `mythplugins-0.20`. There was a small bug found just after the release of MythTV 0.20, which forced a re-release of just the plug-ins. This version is still referred to as 0.20, though.

This is pretty much like all the other MythTV install tasks we have discussed in this book. The next time you start `mythfrontend` on the machine that you ran `make install` on, the additional plug-in functionality will be available.

Exploring MythTV Plug-Ins

Now let's explore the MythTV plug-ins that are in the official plug-ins download. These plug-ins can be broken up into three categories: informational plug-ins, playback plug-ins, and others. We'll discuss each in turn.

Informational Plug-Ins

The first group of plug-ins we will discuss are the informational plug-ins. These plug-ins all provide information in the MythTV user interface that would otherwise be absent. None of this information is really about TV, but it's all useful—these are the things that create that "mythical convergence box" that provided the name (and the database name mythconverg, in case you wondered) for MythTV. Figure 9-1 shows what the Information Center menu looks like in MythTV.

Figure 9-1. *The MythTV Information Center menu*

All the menu items shown in Figure 9-1 are informational. All but one of them is a plug-in as well. The exception is the System Status menu item, which is built into MythTV and shows useful information about the current state of the MythTV backend.

Phone (MythPhone)

The MythPhone plug-in allows you to make and receive Voice over IP (VOIP) Internet telephone calls on your MythTV machine, if you have a microphone installed. We discuss this more in its own chapter, Chapter 14.

News Feeds (MythNews)

MythTV ships with a web syndication aggregator. Many websites offer their content in computer-parseable syndication formats such as RSS and ATOM. These syndication formats are taken by applications called *aggregators* and displayed to the user as the sites change. For MythTV, MythNews is this aggregator, and you can use it to display updates on your TV set. MythNews comes with a large selection of preconfigured sites, which you can find in Utilities / Setup ➤ Setup ➤ Info Center Settings ➤ News Settings (see Figure 9-2).

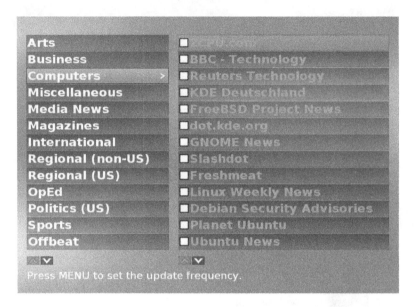

Figure 9-2. *MythNews offers a large number of sites preconfigured.*

Once you've selected the feeds to receive, you can view them in Information Center ➤ News Feeds. The interface looks like Figure 9-3.

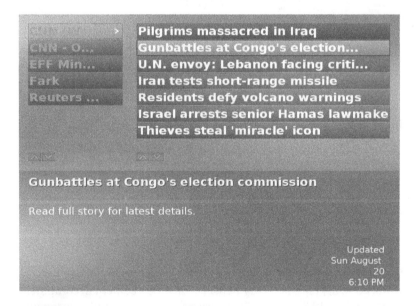

Figure 9-3. *The news display interface*

If you want to go to the web page that a news item refers to, hit Enter, and MythBrowser will launch. To subscribe to other sites that are not on the MythNews default list, press the M key or press Menu on your remote in the news display interface, and you can add and edit subscriptions there.

Weather (MythWeather)

MythWeather is a plug-in that displays weather information for selected cities. At the time of writing this, MythWeather supports a whopping 38,123 locations around the world via Weather.com's online weather service. First you need to configure which location you are in and whether you want metric or imperial units for display. Then you need to configure how "aggressive" you want MythWeather to be. This aggressiveness setting decides how long to wait before deciding that a connection to Weather.com has timed out; the value for you will vary depending on how fast your Internet connection is.

The configuration screen appears if you either go to Utilities / Setup ➤ Setup ➤ Info Center Settings ➤ Weather Settings or select Information Center ➤ Weather without having ever configured MythWeather. Figure 9-4 shows an example of the configuration screen.

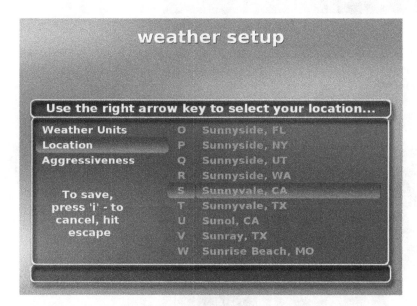

Figure 9-4. *Configuring MythWeather*

When you run MythWeather from Information Center ➤ Weather, the current weather is downloaded from the Internet. This can take a little bit of time, so you will probably end up with the screen shown in Figure 9-5.

Figure 9-5. *Fourth weather screen. Dopplar Radar*

Once everything is loaded, you end up cycling through these three screens (see Figures 9-6 through 9-8). You can manually advance through them using the arrow keys as well.

Figure 9-6. *First weather screen, Current Conditions*

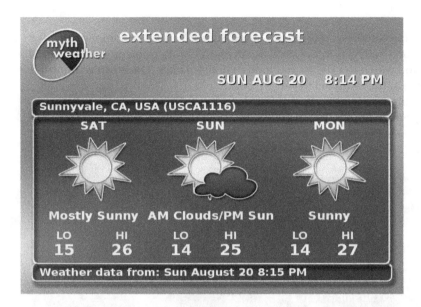

Figure 9-7. *Second weather screen, Extended Forecast*

Figure 9-8. *Third weather screen, Tomorrow's Outlook*

Netflix (MythFlix)

MythFlix integrates with the Netflix service in the United States. Netflix is a company that for a subscription fee will send you DVDs in the mail, much like a library implemented on top of the U.S. Postal Service. You watch the DVDs, and then mail them back. Once you've returned DVDs, you can have more of them mailed to you.

First you need to configure MythFlix. You do this by going to Utilities / Setup ➤ Setup ➤ Info Center Settings ➤ Netflix Settings. Here you will be asked to list the Netflix RSS feeds in which you are interested (see Figure 9-9). You can learn more about RSS feeds and subscriptions in the "News Feeds (MythNews)" section earlier.

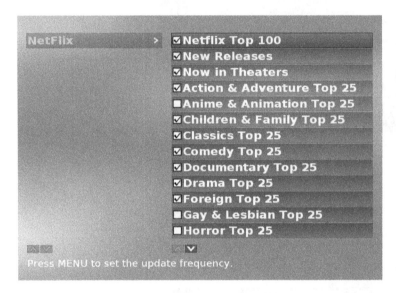

Figure 9-9. *Setting up MythFlix*

Now you can use the Netflix items under Information Center ➤ Netflix. The browse interface lets you see what movies have been added to the Netflix library recently. For example, the remake of *Charlie and the Chocolate Factory* is now available (see Figure 9-10).

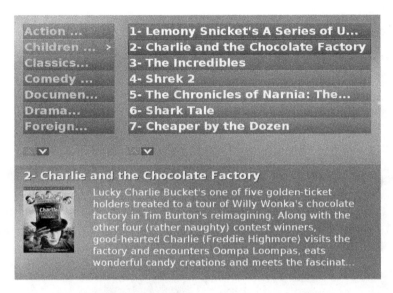

Figure 9-10. *Browsing for movies*

MythFlix also allows you to see the DVDs in your queue (the list of DVDs that you would like shipped to you next), as well as add to this queue based on your browsing. For this to work, you need to configure your account; unfortunately, there currently isn't a very good user interface for this. For more information about how to extract your customer ID from your browser cookie and how to execute the SQL to set up your account, check out the excellent tutorial at http://wiki.knoppmyth.net/index.php?page=MythFlix.

Web (MythBrowser)

MythBrowser is the web browser plug-in for MythTV. It's easiest to start using MythBrowser by selecting a bookmark, although it is also possible to enter a URL at any time by hitting the I key. These bookmarks are created in Utilities / Setup ➤ Setup ➤ Info Center Settings ➤ Web Settings. When you first use MythBrowser, you'll see the blank screen shown in Figure 9-11.

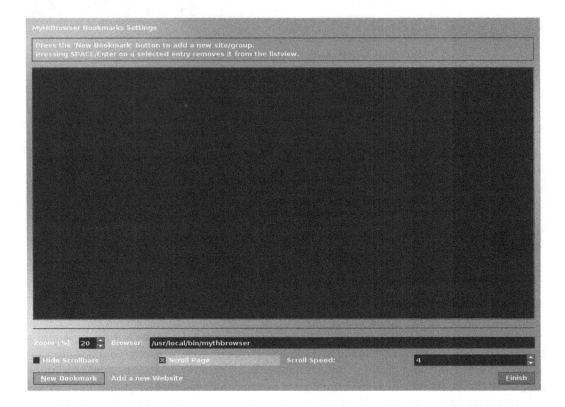

Figure 9-11. *The default MythBrowser setup has no bookmarks.*

You set up bookmarks, which are effectively the starting points of your use of the browser by selecting the New Bookmark button at the bottom of the screen. Once you've set up some default bookmarks, you will end up with a tree view like the one shown in Figure 9-12.

When you're done, select Finish. You can now start the browser by going to Information Center ➤ Web. Here you will see a collapsed view of the tree from the setup screen. Select an entry from the tree to expand it, and then select one of the subitems. Figure 9-13 shows what the MythTV website, mythtv.org, looks like on a TV.

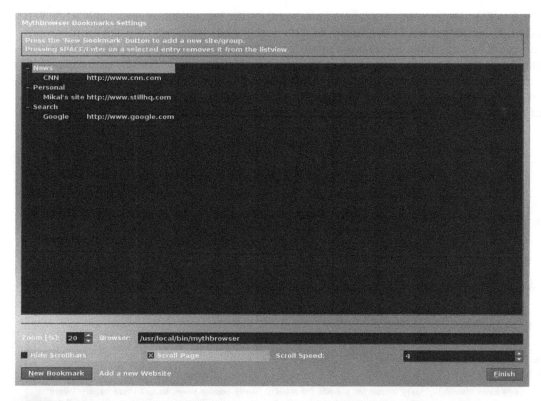

Figure 9-12. *With some bookmarks*

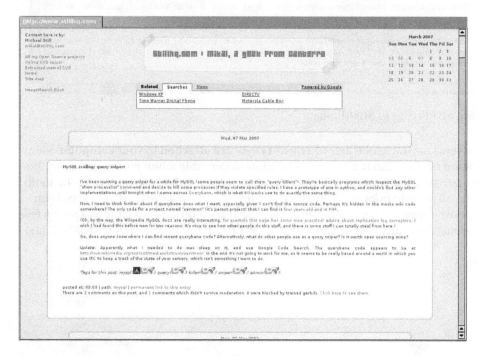

Figure 9-13. www.mythtv.org *in MythBrowser*

You'll notice here that the font is quite small for the text displayed, especially when you take into account that a TV is going to be a lot farther away from you than a computer monitor would be (and that a TV set has a much lower resolution, usually, than a PC monitor). You can fix this by increasing the zooming factor in the MythBrowser settings, with the possible range of zooms being from 20 percent to 300 percent. Our installation defaulted to a zoom of 20 percent, which is way too small. Figure 9-14 shows an example of mythtv.org with a zoom of 150 percent.

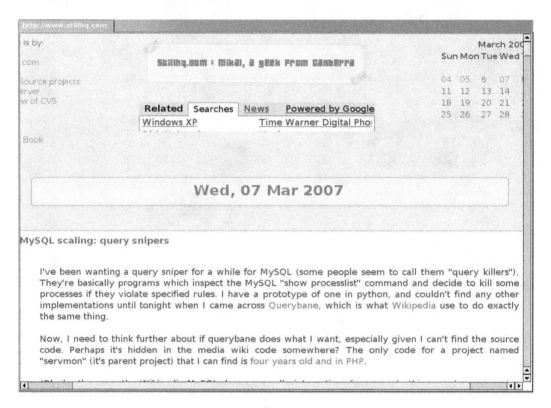

Figure 9-14. www.mythtv.org *in MythBrowser*

MythBrowser is most useful when you have a keyboard attached to your computer. Many wireless keyboard options can make this much more acceptable when you're sitting on the couch on the other side of the room from the computer, such as radio keyboards and Bluetooth keyboards. You can also get infrared keyboards, but you may have to be careful to avoid conflicts with remote controls and, possibly, IR data links on camcorders and laptops. Quite a few keyboard shortcuts are available, as shown in Table 9-1 (from the MythBrowser README file).

Table 9-1. *MythBrowser Keys*

Key	Action
Q	Goes to previous link (eventually in the previous frame)
Z	Goes to next link (eventually in the next frame)
Esc	Exits
Space/Return	Follows selected link
R, Backspace	Goes back to previous page
Cursor Up/Down	Scrolls
Cursor Left/Right	Scrolls
Page Up/Down	Scrolls
P	Switches to next tab
D	Deletes current tab
I	Shows the Enter URL dialog box
M	Shows a pop-up menu
<	Zooms out
>	Zooms in
F1	Toggles where keyboard input goes
2	Moves mouse pointer up
8	Moves mouse pointer down
4	Moves mouse pointer left
6	Moves mouse pointer right
5	Acts as a left mouse button click

The most important key to note here is F1, which toggles between using the keyboard input to drive MythBrowser and allowing you to enter text into text boxes on web pages.

System Status

As mentioned earlier, the System Status menu item is built into MythTV and technically isn't a plug-in. However, this is a distinction at the code level and is not apparent in the user interface. The System Status screen tells you important information about the current state of your MythTV machine.

Figure 9-15 shows the first screen of the System Status section. This screen shows you when the guide data was last downloaded by mythfilldatabase and, more important, when you will run out of guide data.

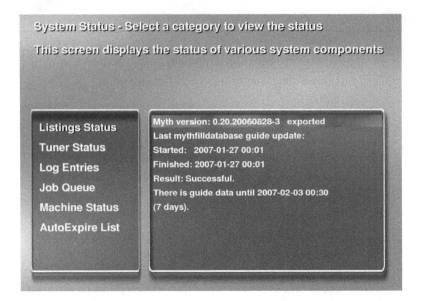

Figure 9-15. *The first page of System Status*

You can see that down the left side there are a number of other options for more information about the current status of the MythTV system. The tuner status looks like the one shown
in Figure 9-16, with more detail for a given tuner being available if you select the tuner (see
Figure 9-17).

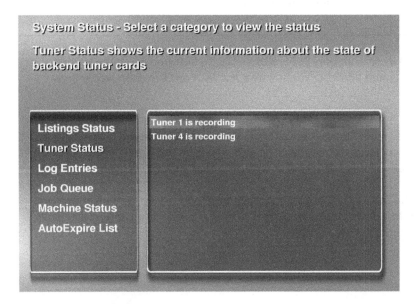

Figure 9-16. *Tuner status*

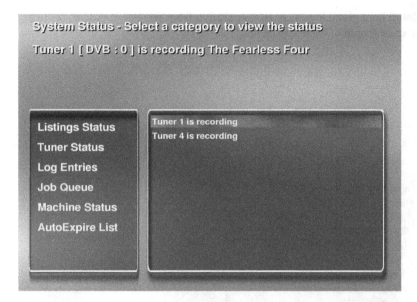

Figure 9-17. *Detailed tuner status*

Log entries help explain what the system has been doing. For example, you can see in Figure 9-18 that shows have recently been expired, that commercial flagging has been running, and that one show has been recorded recently. You need to have logging enabled; go to Utilities / Setup ➤ Setup ➤ General and then go to the fifth screen (Myth Database Logging).

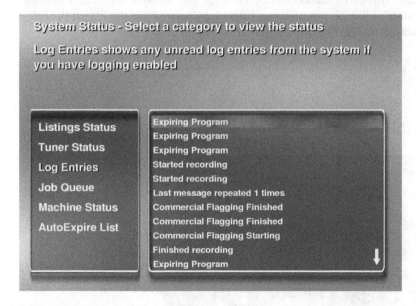

Figure 9-18. *Sample log entries*

You can see more detail on a given log entry by selecting it from the list of entries on the right. Figure 9-19 shows an example of the extra detail.

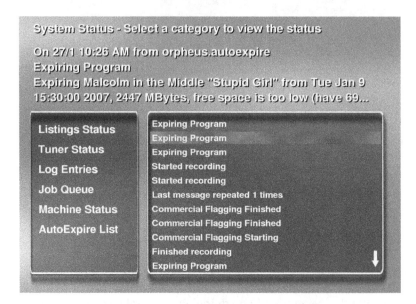

Figure 9-19. *Detailed log entry for a show expiry*

The MythTV job queue is the list of tasks that the MythTV backend has queued for processing at the current time. This can include commercial flagging, transcoding, and user jobs. You can see the current state of the job queue if you select Job Queue from the list on the left. You will see a list of outstanding jobs like the one shown in Figure 9-20.

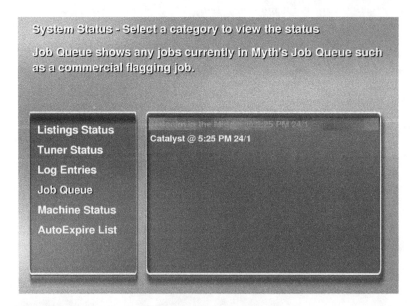

Figure 9-20. *Jobs currently in the backend queue*

Once again, selecting a job will show more detail, as shown in Figure 9-21.

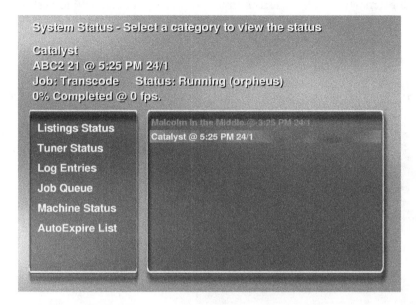

Figure 9-21. *More detail from a transcode job in the backend queue*

Machine Status will show you the status for each of the machines in your MythTV setup. This is quite handy, because it will report on remote machines as well as the one you are currently running the MythTV frontend on. Figure 9-22 shows an example.

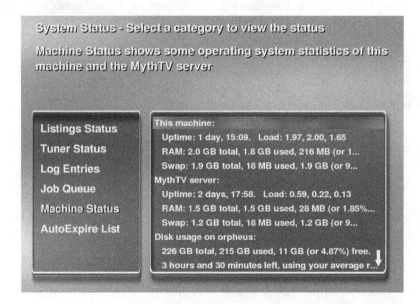

Figure 9-22. *Machine status*

Once more, you can see more detail by selecting a row in the list on the right, although selecting rows that describe a machine will cause a complete summary of the machine to be displayed, as demonstrated in Figure 9-23.

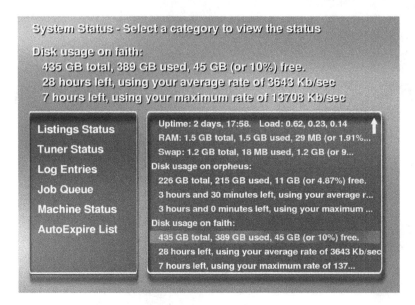

Figure 9-23. *More detail on one machine in our MythTV setup*

Finally, you can see the autoexpire list (the list of recordings which will be automatically deleted next if space needs to be freed up), as shown in Figure 9-24, and additional detail on those recordings, as shown in Figure 9-25.

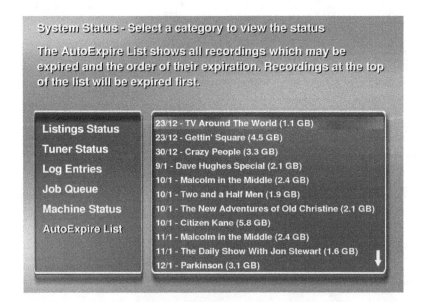

Figure 9-24. *The list of recordings to expire next*

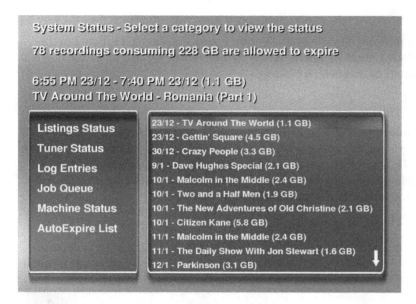

Figure 9-25. *Detail on a soon-to-be-expired recording*

Playback Plug-Ins

The next group of plug-ins we'll discuss are what we're calling the *playback* plug-ins. All these plug-ins implement audiovisual features in addition to those already implemented by MythTV. Figure 9-26 shows the Media Library menu under which all these plug-ins reside.

Figure 9-26. *The Media Library menu*

Watch Recordings

Watch Recordings isn't a plug-in; it's core MythTV functionality. Because it is listed in this menu, we wanted to include a short section here so you didn't think we had forgotten it. We covered watching recordings extensively in Chapters 4, 5, and 6.

Watch Videos (MythVideo)

The MythVideo plug-in allows you to watch video files that weren't recorded by MythTV. It does this by indexing the videos in a specified directory, looking up helpful information about the videos on Amazon's IMDB online movie database, providing a user interface to select videos, and executing a player for selected videos. The first step is to install missing dependencies. For our Ubuntu machine, this meant installing this package:

```
$ sudo apt-get install libxml-simple-perl
```

The next step is of course to configure MythVideo so it knows where to look for those videos. You do this in Utilities / Setup ➤ Setup ➤ Media Settings ➤ Videos Settings ➤ General Settings. The first screen of the configuration looks like Figure 9-27.

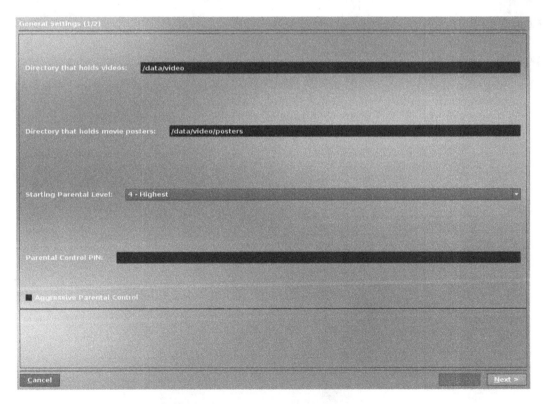

Figure 9-27. *Configuring MythVideo*

You can see here that the first step in configuring MythVideo is to configure the directory that is scanned for videos, the directory that will store the downloaded movie posters for videos, and the parental controls that will be applied to these videos. In the next screen, you configure the browsing interface, which you will use to select videos to watch (see Figure 9-28).

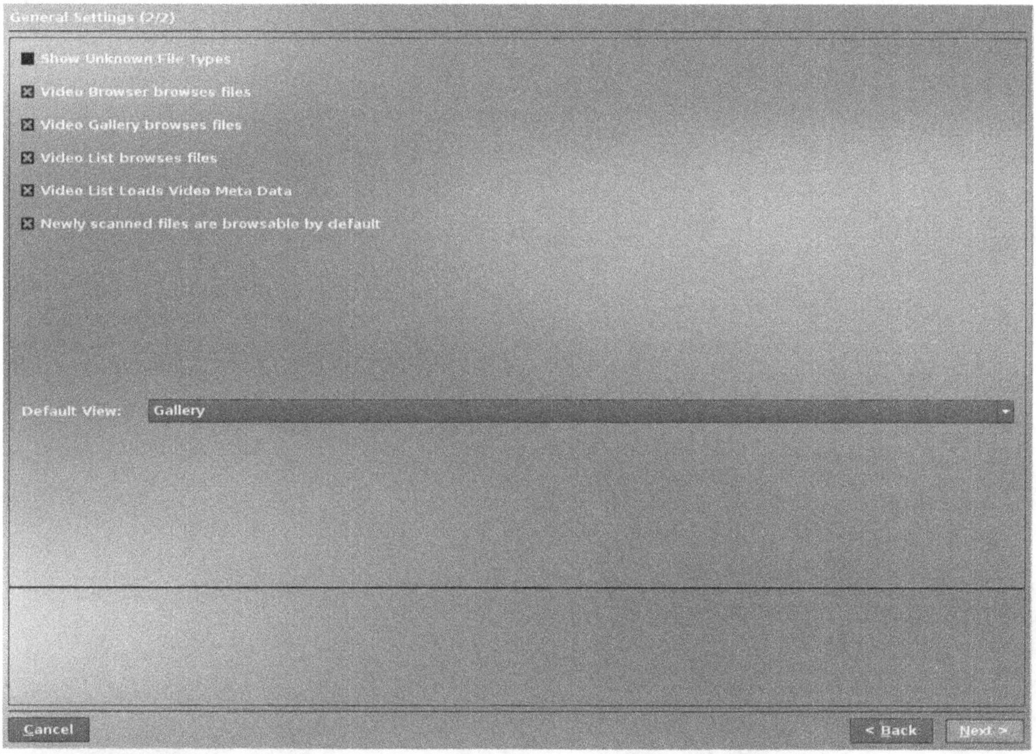

Figure 9-28. *Configuring the MythVideo browser*

Let's stick to the default gallery for now; we will show you what each of the options looks like later. Next you need to configure where MythTV can find the IMDB scripts that are used to download information about the videos from Amazon's IMDB online database, including posters for movies if they are available. The lookups are based on the names of the video files, so these need to match the names of the movies if possible. You can, however, force a given IMDB entry to be mapped to a video, and we will show you how to do that later too. Figure 9-29 shows the IMDB configuration screen.

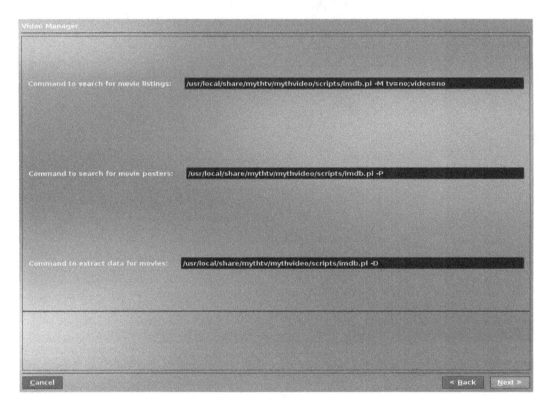

Figure 9-29. *Configuring the MythVideo IMDB lookup*

Finally, Figure 9-30 shows some simple options for the default video browser.

Figure 9-30. *Configuring the MythVideo gallery browser*

Next you need to configure the player to use for the videos. You do that in Utilities / Setup ➤ Setup ➤ Media Settings ➤ Videos Settings ➤ General Settings. This setup screen is so simple that we're not going to show it here. It's a single textbox, which contains the command line for the video player to be executed when a video is selected for playback. The default is as follows:

```
mplayer -fs -zoom -quiet -vo xv %s
```

The %s here will be replaced with the filename of the video to be played back. You can use any video player that you want to here, but you should select something that works well with the lirc remote control system described in Chapter 3. Standard choices are mplayer and Xine. Xine is the player we're going to use for some DVDs in Chapter 12. If you want to use Xine here, then the command line would look like this:

```
xine -pfhq -no-splash %s
```

This will do the same thing as the earlier command line, except with Xine. As you can see, you don't need to specify the full pathname to the player; the frontend will use the Linux search path to find it automatically. However, either way, you must have the player installed on the frontend you're using.

Finally, you can tell MythVideo to use different commands for different file types. You do this in Utilities / Setup ➤ Setup ➤ Media Settings ➤ Videos Settings ➤ File Types, where you will see a screen like the one shown in Figure 9-31 for each of the file types that MythVideo recognizes.

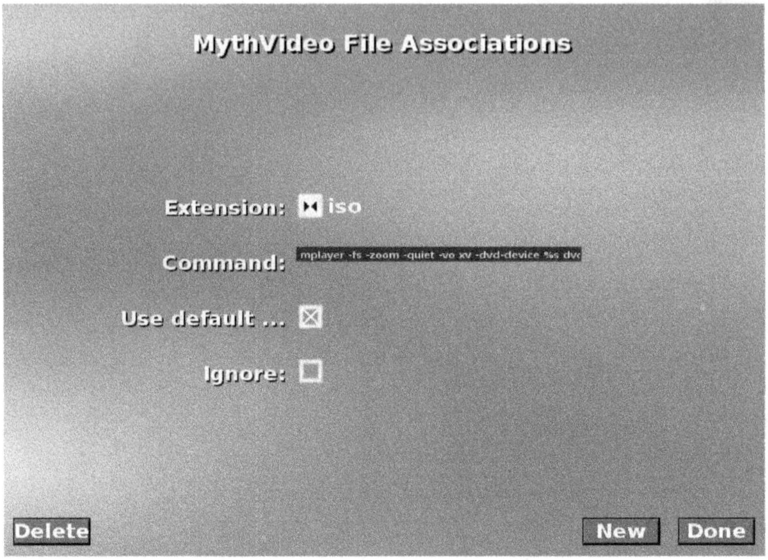

Figure 9-31. *Configuring different players for different file types*

If you select the default player, then you will get the one we configured in the previous screen. You can of course specify a custom player for a given file type, as well as tell MythVideo to ignore that file type entirely, like you would for log files or text files.

Next you need to have MythVideo find all the video files in the specified directory. You do that with the video manager, which is located at Utilities / Setup ➤ Video Manager. You'll see a progress bar as the videos directory is scanned recursively. If you have a lot of files in that directory, then this can take quite some time (see Figure 9-32).

Figure 9-32. *MythVideo's video manager detecting files*

It isn't too bad for us, though; it took only a couple of seconds to detect our videos. Once there, you'll see a list of the videos that were detected. If you press Menu on your remote control, press the M key if you have standard key mappings, or press the right arrow on either your keyboard or remote, then you'll see the menu shown in Figure 9-33.

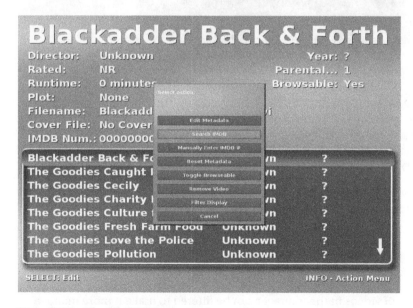

Figure 9-33. *The MythVideo manager menu*

You can see in the figure the video manager's options. Briefly, they are as follows:

Edit Metadata: This option presents you with an editing interface for the metadata for a video. Here you can specify a custom player, edit the title of the video, and change other metadata.

Search IMDB: This option lets you search for the title of the video on Amazon's IMDB. The title is simply defined as the filename of the video, with some simple transformations (for example, underscores are replaced with spaces). Once you've performed this lookup, you end up with full metadata with little effort. Figure 9-34 shows an example.

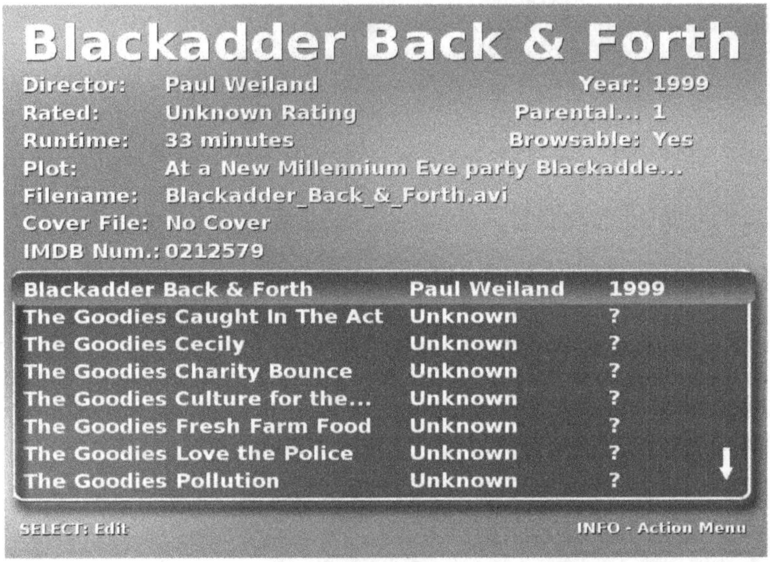

Figure 9-34. *The MythVideo IMDB lookup provides metadata.*

Manually Enter IMDB #: If the title lookup fails, then you can use this option to enter the IMDB entry number of a movie. You find out the IMDB entry number by searching on the http://www.imdb.com website for the movie and then looking for the number in the URL. As an example, *The Whole Nine Yards* has this URL on IMDB: http://www.imdb.com/title/tt0190138/. The IMDB number in this case is 0190138.

Reset Metadata: If you download metadata for a video and then decide that it's wrong, you can use this option to reset it to the defaults.

Toggle Browseable: This will change whether this video appears in the video selection user interface.

Remove Video: This will delete the video from the hard disk.

Filter Display: The list of videos in the manager can be filtered to make it more manageable. Select this option, and you will see the screen shown in Figure 9-35.

Cancel: Finally, you can just exit the menu.

Filters video list
Result of this filter : 18 video(s)

Category : ▶◀ All

Genre : ▶◀ All

Country : ▶◀ All

Year : ▶◀ All

Runtime : ▶◀ All

User Rating : ▶◀ All

Browse : ▶◀ All

Sort by : ▶◀ title

Done

Save as default

Figure 9-35. *Filtering options for the MythVideo video manager display*

At last you are ready to watch some videos. You do this with the MythVideo user interface, which is located at Media Library ➤ Watch Videos. You'll see the gallery interface by default, which looks like Figure 9-36.

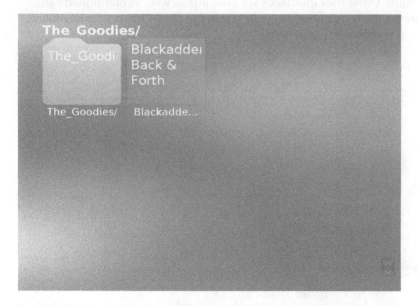

The Goodies/

The_Goodi   Blackadder
             Back &
             Forth

The_Goodies/   Blackadde...

Figure 9-36. *The MythVideo gallery interface*

If you select a folder, then the interface will look like Figure 9-37.

Figure 9-37. *The MythVideo gallery interface inside a subfolder*

The item in the top left will return you to the parent folder. To play a video, simply select it in the interface, and press the OK button on your remote or the Enter key on your keyboard. This will launch the player specified for this file type (or the specific player for this video if you have overridden the default). Other user interfaces are available as well, as mentioned earlier in this section. If you press the Menu button on your remote or the M key on the keyboard, then you will see the options shown in Figure 9-38.

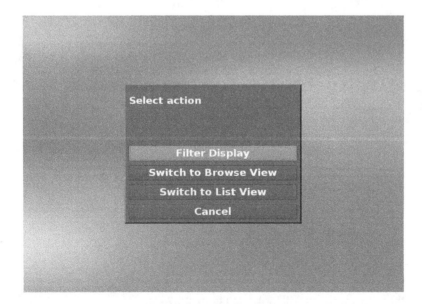

Figure 9-38. *Other video selection options*

Another option is the browse view, which is great if you have a relatively small number of videos or want to see a lot of detail about a video (see Figure 9-39).

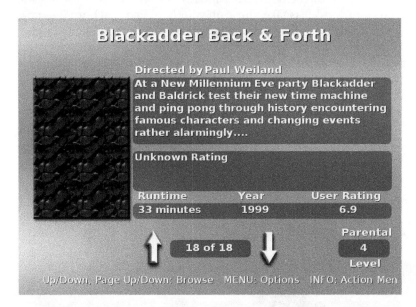

Figure 9-39. *Browse view*

Finally, the list view works much better with large numbers of views in groups (and emulates the selection user interface used for TV programs), as shown in Figure 9-40.

Figure 9-40. *List view*

Listen to Music (MythMusic)

MythMusic is a player for your audio files, including MP3 and AAC. It can also rip CDs and add them to your collection if you want, as well as take music off your hard disk and burn it to new CDs. Lots of people love MythMusic, and some people hate it. If you're not too keen on the look of MythMusic based on what you read in this section or decide you don't like it after using it for a bit, then you might consider trying the SoftSqueeze plug-in instead, which unfortunately we don't cover in this chapter.

First, you configure MythMusic by visiting Utilities / Setup ➤ Setup ➤ Media Settings ➤ Music Settings. Here you will see general settings, player settings, and ripper settings. The general settings cover features such as where the audio files are stored, which device file to use for your CD drive, how to sort the music in your collection, and what filename format to use for ripped music (see Figure 9-41).

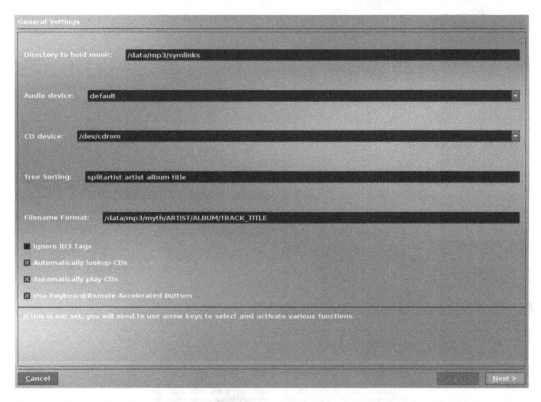

Figure 9-41. *Configuring MythMusic's general settings*

The next screen of the general configuration options covers burning new CDs. You select the CD drive to use to burn these CDs, the size of the disk that the music will be burned to, and some aspects of the CD-burning process such as how to blank rewritable CDs (see Figure 9-42).

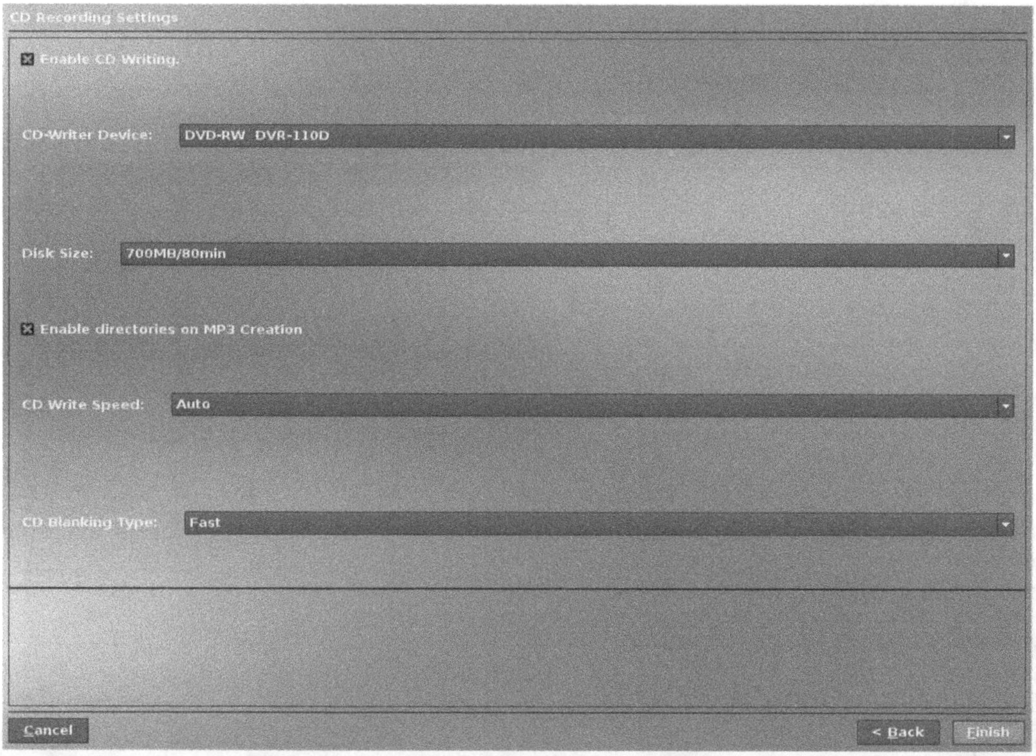

Figure 9-42. *Configuring more of MythMusic's general settings*

Next you need to set up the music player. You do this in Utilities / Setup ➤ Setup ➤ Media Settings ➤ Music Settings ➤ Player Settings. First, you get to select the type of shuffle that should be used. The available options are as follows:

Normal Shuffle: The tracks are in the order they were found on disk. This will equate to being ordered by artist, album, and then track on a given album.

Random Shuffle: The tracks are in a random order.

Intelligent Shuffle: The ordering takes into account the rating of the track, the number of times the track has been played, how long ago the last play was, and then a random factor.

Each of the factors for the intelligent shuffle allow for weightings as well, which allows you to specify the relative importance of the factors listed in the last bullet point. The higher the number as a weighting, the more attention is paid to that factor. Figure 9-43 shows the configuration dialog box.

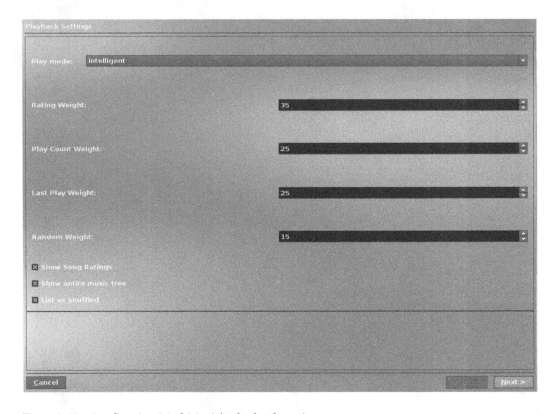

Figure 9-43. *Configuring MythMusic's playback settings*

Next are the settings for the visualizations. Visualizations are graphical displays that run during the playback of a track. Figure 9-44 shows an example of a small visualization in this playback.

Figure 9-44. *A sample visualization*

You configure the options for visualization on the next screen in the player setup wizard, which looks like Figure 9-45.

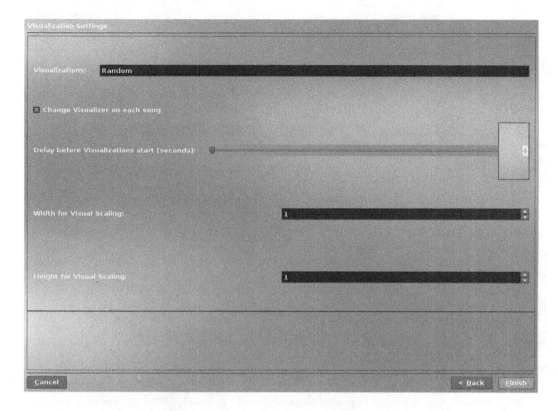

Figure 9-45. *Setting up visualizations*

Finally for the setup, you might want to configure the ripper settings. The ripper is the software component that takes CDs and turns them into audio files on your hard disk. You can first configure the paranoia level of the ripper (how hard the ripper works to make sure that the data it is reading is valid), the naming convention for files, and a couple of import options (see Figure 9-46).

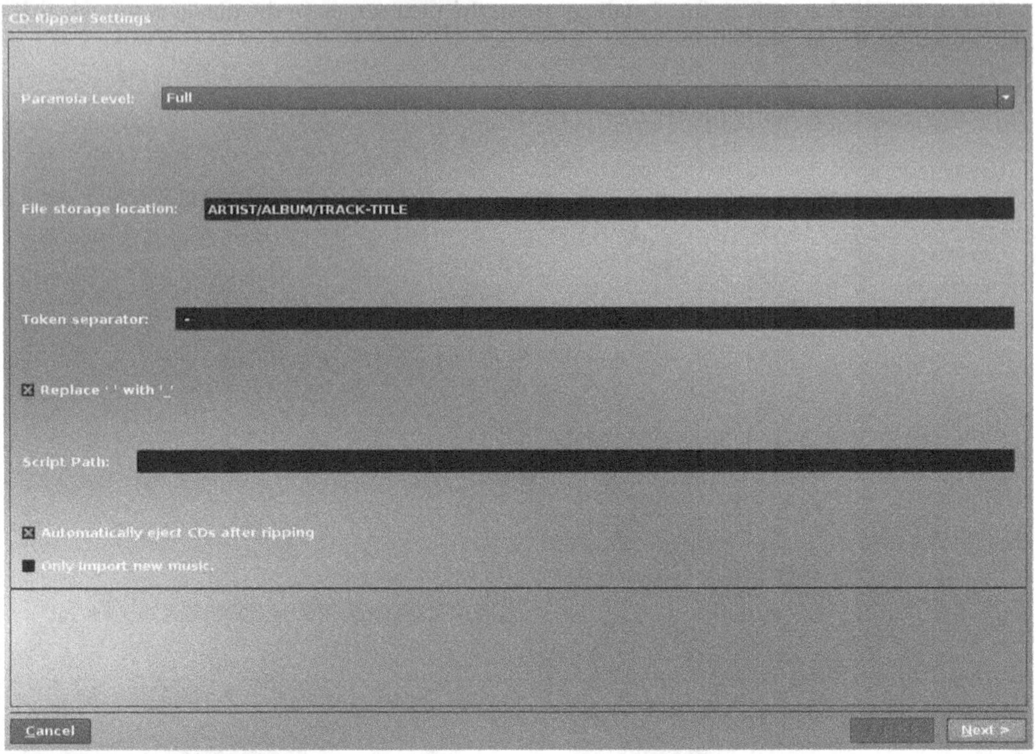

Figure 9-46. *Ripper settings*

A few more ripper settings are available as well, such as which MP3 encoder to use and what quality MP3s to produce from the ripper (see Figure 9-47).

After all that configuration, you're still not ready to use MythMusic. If you already have music on the hard disk, then you need to scan for that music. Do that by going to Utilities / Setup ➤ Music Tools. Here you will find tools that help you set up the music library to your satisfaction (see Figure 9-48).

CD Ripper Settings (part 2)

Encoding: Lame (MP3)

Default Rip Quality: High

☒ Use variable bitrates

Cancel   < Back   Finish

Figure 9-47. *More ripper settings*

If you click the Cancel button to start out, you won't want to add information. To
keep using the Import Music option, use a rip up the CD to audio files.

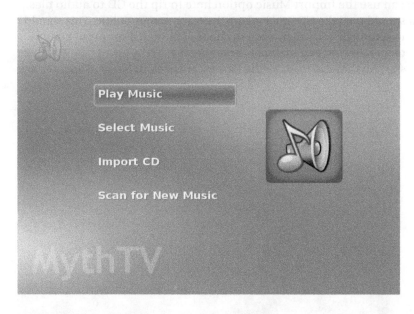

Figure 9-48. *The MythMusic music tools*

If you have music on the hard disk, then you need to select the Scan for New Music option. This will pop up a progress bar and might run for quite some time if you have a lot of music on the disk. The progress bar will bounce backward and forward until the operation is done (see Figure 9-49).

Figure 9-49. *MythMusic scans for new music, which can take quite some time.*

If you don't have any music in your collection to scan or if you want to add more music to the collection, then you can use the Import Music option here to rip the CD to audio files. When you insert a CD, it should read the CD and look up the track names online. This should work almost all of the time, although you might need to choose between a couple of albums. Worst case, if the track lookup fails, you can manually enter details.

Now you're ready to start MythMusic from Media Library ➤ Listen to Music. The interface looks like Figure 9-50.

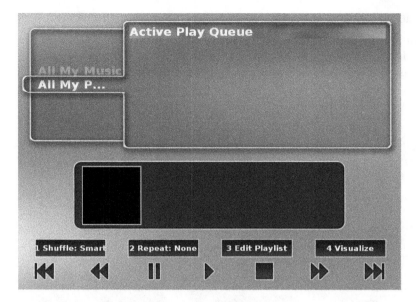

Figure 9-50. *The MythMusic playback interface*

The default starting place is the playlist interface, which isn't very exciting for us, because we don't have any playlists configured. You can select by artist, however, which you select from the list on the left. You can then select songs by artist (see Figure 9-51).

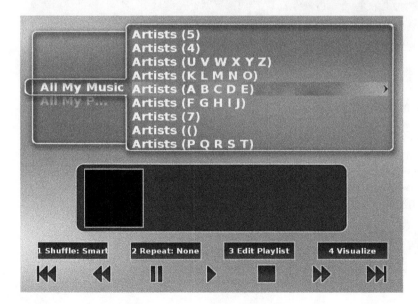

Figure 9-51. *The MythMusic playback interface, selecting by artist*

You can of course play a randomly shuffled selection from your entire collection.

Image Gallery (MythGallery)

MythGallery displays your photo collection as a slide show on your TV set. You need to configure where on your disk photos are stored. You do this via Utilities / Setup ➤ Setup ➤ Media Settings ➤ Images Settings. You'll get a configuration screen like Figure 9-52.

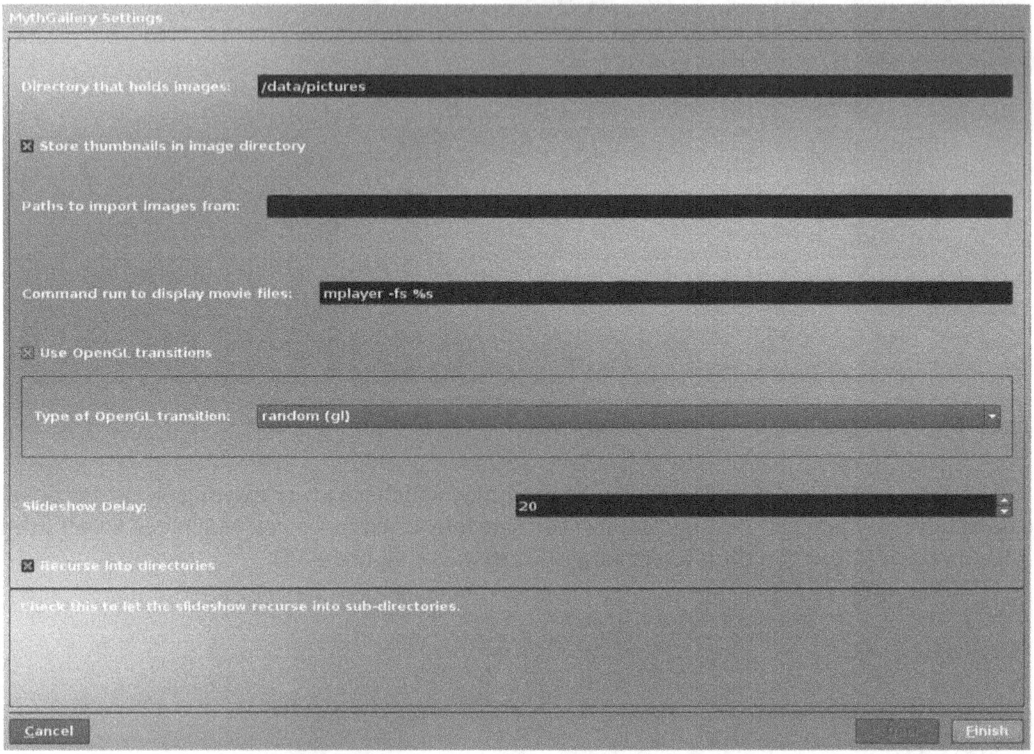

Figure 9-52. *Configuring MythGallery*

The most important options here are where to look for the pictures, how long to display each picture in a slide show, and whether to recurse into directories (which should almost always be enabled). Once you have MythGallery configured to your satisfaction, you run it by selecting Media Library ➤ Image Gallery. You'll see a user interface in which you can select a folder to view images from, which looks like Figure 9-53.

Figure 9-53. *Selecting a folder to display*

Press either the Menu button or the M key to get to the menu on the left of the screen. In that menu you can select which folder to display images from (or randomly display from all images), and so forth. You can also rotate images from this menu as well.

Play Games (MythGame)

MythGame is a games console emulator for your MythTV system. It's quite different from the rest of MythTV in that it needs a lot of configuration on its own and is a little complicated. The basic concept is that you install players, which emulate a given game console, and then install games for that player. A large number of players are supported, including SNES, NES, Game Boy and Game Boy Advance, Nintendo 64, Amiga, Lynx, Sega Mega Drive and Genesis, PlayStation, and more. You can also purchase interesting USB controllers to help with your gaming experience. Make sure you check that the controller you are thinking of buying works with Linux before you make the purchase, though.

For more information about setting up MythGame, you should check out the MythTV documentation at http://www.mythtv.org/wiki/index.php/MythGame and http://www.mythtv.org/wiki/index.php/MythGameEmulationSetup. We won't cover any more about MythGame here, because the excellent documentation on those pages is much less likely to become out-of-date.

Other

A few other plug-ins don't fit into the previous categories, so we'll discuss them separately in the following sections.

MythDVD (Optical Disks)

MythDVD is the DVD playback plug-in, which is discussed in its own chapter, Chapter 12.

MythControls

MythControls is a user-friendly way of changing which keys on the keyboard map to commands in the MythTV user interface. It's also fairly new, because it was introduced in MythTV 0.19. Changing keyboard bindings might make sense if you have an unusual keyboard layout or find the default setup annoying. You can find MythControls under Utilities / Setup ➤ Edit Keys, and it looks like Figure 9-54.

Figure 9-54. *MythControls lets you edit the keyboard mappings for commands.*

For example, in the Playback options, we hate how pause is P. We would much rather that it was the spacebar like it is in Xine. So, if you select TV Playback on the left and then scroll down to select Pause, you can edit the key that is mapped to this command (see Figure 9-55).

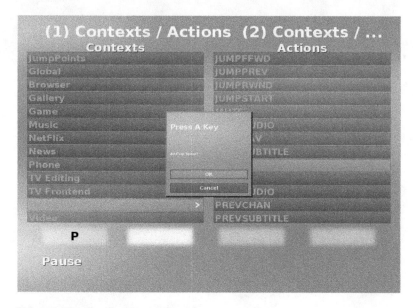

Figure 9-55. *Editing the keyboard mapping for pause*

The extra input fields are because you can have more than one key mapped to a given behavior. To change the value in an input field, select that field, then hit Enter (OK on the remote), and finally press the key sequence you would like to use. You will be prompted to see whether it is correct (see Figure 9-56).

Figure 9-56. *Confirming a change*

Conclusion

In this chapter, you took a quick tour of the various plug-ins that come with MythTV. These plug-ins perform a variety of useful tasks, as well as provide an extensible mechanism for you to implement your own MythTV functionality. Read on, because in the next two chapters we will discuss how to extend your MythTV box once you become addicted and how to install and use the web interface, which lets you do cool things like schedule recordings while you're out and about.

CHAPTER 10

■■■

Expanding MythTV

You can expand your MythTV setup in four ways. First, now that you've had MythTV for a while, you've probably discovered that a lot of shows on TV are worth watching. Not only that, but if you've introduced anybody else to MythTV, you've probably discovered that sometimes more than one program that you want to record is on at once. This is called a *conflict*. Luckily, it's relatively easy to add more TV tuners and more disks to a MythTV box to avoid this problem; we'll show you how in this chapter.

Second, you might be running low on disk space by now. This problem will only get worse if you add more tuners, because your television backlog will grow as you start to capture more content. We'll show you how to add more disks to your MythTV system in this chapter as well.

Third, if you are doing a lot of transcoding, then you might notice that the backend machine is starting to get more and more behind. This is especially true if you're transcoding recordings for your Sony PlayStation Portable or Apple iPod, both of which are relatively expensive transcoding operations. We will show you how to add another backend to your MythTV installation, which means you can run some of those transcoding jobs on a second machine. This also has the advantage that you can add tuners to the second backend machine if you are out of expansion space on the first backend.

Finally, you can add more frontends so you can watch more than one recording at once. We won't cover this in this chapter, because Chapter 8 was devoted to this topic.

Note Some of the procedures described in this chapter are quite safe for new Linux users, whereas others run the risk of removing all your data. Forever. We'll let you know when we're about to discuss something risky so you can make the appropriate backups beforehand.

Avoiding Expanding

As we just mentioned, some of the procedures in this chapter are not for the faint of heart, especially some ways of adding more disk space for TV recordings (until MythTV 0.21, that is). If you're not comfortable with doing something, don't. It is perfectly OK to take the easy way out and wait for the next release (or run the latest development release from SVN if you're feeling a little adventurous; see Chapter 15 for more information about this) or just replace a part and reinstall it than to work too hard on expanding something.

Adding More Tuners

Although MythTV will attempt to resolve program conflicts by changing the time of day that it records shows at, you will sometimes still miss out on shows. An example is when a show airs only once and that time conflicts with a show with a much higher priority or that also airs only once. To record more than one show at once (or overcome the problem of having to switch channels to record the next show and missing the end of the first show), you need more than one TV tuner. Most people start with one tuner and then add another later. Before adding extra tuners, you need to consider a few issues, such as determining whether you have enough disk bandwidth to record more than one program at once and determining how much more disk space you are going to need. The following sections will address these issues.

Determining Your Disk Bandwidth

Unless you are recording several simultaneous high-definition channels or very poor quality analog (that isn't compressing well), disk bandwidth (how much you can read or write to disk at a time) shouldn't be a problem. However, the more tuners you have in the one machine, the more available RAM you'll want for caching data before it's written to disk. Otherwise, the operating system will be forced to write smaller portions of each program to disk, which causes disk seeks (and lower performance) or file fragmentation (which will also degrade how fast you can read the data later). One gigabyte of memory should be enough for a machine running the backend, the frontend, and a few tuners.

Using RAID 0, also known as *disk striping*, will increase available bandwidth if you need it. Before you can determine whether your system is struggling with its IO load, you need to determine what the disks you are using are capable of doing. We recommend doing this by running a simple benchmark on your disk array when it is idle and then comparing that to the workload when the machine is busy recording the maximum number of streams it is capable of recording. For example, here is a benchmark run on one of our MythTV disk arrays when it is not recording (/data/temp is a location on the disk array that you want to test):

```
$ sudo apt-get install postmark
$ sudo postmark
PostMark v1.51 : 8/14/01
pm>set size 1000 9000
pm>set number 2000
pm>set transactions 5000
pm>set location /data/temp/
pm>set report verbose
pm>run
Creating files...Done
Performing transactions..........Done
Deleting files...Done
Time:
        2 seconds total
        1 seconds of transactions (5000 per second)
```

```
Files:
        4512 created (2256 per second)
                Creation alone: 2000 files (2000 per second)
                Mixed with transactions: 2512 files (2512 per second)
        2483 read (2483 per second)
        2513 appended (2513 per second)
        4512 deleted (2256 per second)
                Deletion alone: 2024 files (2024 per second)
                Mixed with transactions: 2488 files (2488 per second)

Data:
        13.57 megabytes read (6.79 megabytes per second)
        25.36 megabytes written (12.68 megabytes per second)
pm>quit
```

You should note that this benchmark is not a perfect simulation of the actual disk usage pattern that MythTV will create, but it is a compromise that will run relatively fast and produce enough information for you to get some numbers. You can see in the output, shown next, that the approximate write bandwidth of the disk array is 12 megabytes per second. Once you have maximum numbers from this benchmark, you can compare that with the output of iostat to determine how close to the current maximum performance you are. You should do this while your backend is recording the maximum number of streams possible. Here is the output of iostat for one of our disk arrays:

```
$ iostat -dm
Linux 2.6.15-20-686 (molokai)    10/12/2006
```

Device:	tps	MB_read/s	MB_wrtn/s	MB_read	MB_wrtn
...					
md0	29.27	0.16	0.08	678431	365084
md1	46.28	0.27	0.14	1176695	608637
md2	29.55	0.14	0.09	613833	389517

We've included the relevant devices only for the disk array, which is a logical volume spread across md0, md1, and md2. Even though we're currently recording a stream, you can see that we still have lots of disk write capacity, because we're currently using less than a megabyte per second on these disks. Additionally, if you're using IDE disks, then you can use hdparm to tweak the performance of your disks, at the possible expense of the amount of noise that the disks create. You can read more about this at http://www.linuxdevcenter.com/pub/a/linux/2000/06/29/hdparm.html.

Finally, if you are starting to run at the edge of your disk bandwidth, MythTV will start logging warnings, so if you're worried, keep an eye on your log files.

Getting More Disk Bandwidth

If you find you need more disk bandwidth, then you have a few options. Adding more disks to a RAID 0 array can improve its write performance. Buying faster disks (the RPM rating of the disks, which is often 5,400; 7,200; 10,000; or 15,000) will help too. Also, you might find that you need to move to a faster disk connectivity solution; for example, you might change from USB to Serial ATA (SATA). If you have multiple disks, connecting them to separate disk controllers, including different USB controllers, can also help. Many computers have multiple USB ports, and some might share the same controller, limiting the total bandwidth. If you have multiple controllers, balancing the disks between them can improve performance. This is the same with IDE and SATA disks as well. For many consumer-level IDE controllers, the primary and secondary channels share the same bandwidth, and you might have to buy a separate IDE PCI card to get increased disk bandwidth.

Adding an Internal or External Tuner

When adding a new tuner card, you'll also need to decide whether to go for an internal or external card. If adding an internal card, you'll need a free slot inside your machine. For small computer cases (for example, the Shuttle cases), which might have only one PCI slot, you won't be able to add more internal tuners. In this case, you can add either external USB tuners or FireWire tuners. You might also choose to add internal (or external) tuners to another backend that has more PCI slots and is placed somewhere where extra noise isn't too much of a problem.

Remember, if you add a USB tuner, then you need to have enough free bandwidth on the USB bus for the video the tuner is going to stream to the machine. There isn't a simple way to test for this beforehand, so you are best off just plugging a tuner in and seeing whether it works. The kernel will log to `dmesg` if there isn't enough USB bandwidth. If you find you need more bandwidth, try rearranging the USB devices on your machine so that the tuner is on a controller by itself (most modern machines have more than one USB controller). You can find out what is plugged into the controllers by using `lsusb -t`.

You might also have to consider the USB devices' power requirements. A limited amount of power can be drawn through USB ports, and several devices linked to the one USB port (or sometimes several ports on the one controller) can easily exhaust this. If you get error messages in `dmesg` or devices being recognized only one at a time, you might need a powered USB hub to provide enough power.

Configuring the New Tuner for MythTV

Once you have plugged a new TV tuner into your machine, you need to configure MythTV to recognize this new tuner. The process for this is similar to what you went through when you initially set up MythTV with the `mythtv-setup` program. You can read more details about how to set up a tuner in Chapter 4, although you will also need to set up the drivers for your new video capture card, as discussed in Chapter 2. After adding the new tuner, you'll be able to record more than one program at a time as well as set priorities on which tuner card a program should record from (such as if one tuner card is of higher quality than the other).

Adding More Disk Space

As you know, it's possible to store and access more than just TV recordings with MythTV. The media for the video, photo, and music modules all use disk space—possibly quite a bit if you have many videos, photos, or high-quality music.

The cheapest and easiest way of adding more storage is to buy another hard disk and plug it into your MythTV system—either an internal or external disk. An easy way to set it up is to format it as a Linux file system (using the same criteria to select one as you did during the install) and allow (or set up) Ubuntu to mount it on start-up. If you have lots of videos, photos, or music, it might be easiest to add another disk and store some or all of these files on the new disk, away from your TV recordings.

In MythTV 0.20, TV recordings can be stored only in a single directory. MythTV 0.21 (not released at the time of this writing, although the code is in the MythTV SVN repository) will add a feature called Storage Groups. With this feature, you can have different storage groups for different recordings, and each storage group is a list of one or more directories of where to store the recordings. With this feature, you can easily just add more disks to your MythTV setup, individually format them, mount them (without worrying about RAID or LVM), and add them to a storage group (perhaps the default storage group).

However, if you want more space for TV recordings and you cannot wait for MythTV 0.21 (and don't want to run MythTV from the SVN tree), you have two options. You can add a new file system with the new disk and then manually move recordings to it (which has flaws), or you can make the file system that stores your TV recordings larger. You have two ways to do this: you can replace your current disk with a larger one, or you can use LVM or RAID to combine two (or more) disks into one larger one. We will discuss all three processes next.

Adding Another Separate Disk

MythTV will allow you to move recordings from its main recording directory to another directory. You create a new file system on the new disk and mount it somewhere on your system. The process is a little complicated, and if you don't have experience with partitioning and formatting a hard disk, then you're best off seeking the assistance of someone with such experience, because it is quite easy to corrupt all your data in the process.

Once you have a new partition to store video on, it's simply a case of moving video files to the new directory and then creating symlinks from the old location to the new one. For example, to move a recording from /var/video to /var/video2, you might use these commands:

```
$ mv /var/video/1002_20070121213000.mpg /var/video2/
$ ln -s /var/video2/1002_20070121213000.mpg /var/video/1002_20070121213000.mpg
```

If you enable the Follow Symbolic Links When Deleting Files option from the third screen of the general options configured with mythtv-setup, then even the file in the new location will be deleted when the recording is removed via the MythTV frontend or MythWeb.

Replacing Your Current Disk

If you just want to replace your hard disk in an existing machine with a larger one (perhaps you don't want the extra heat and noise from having two disks), you can simply copy all your data from the old to the new and then swap them over.

Caution It is possible to lose all your data with the steps described in this section. Either be sure of what you are doing or ask someone with more experience to guide you through the process. The commands we talk about in this section do *not* ask "Are you sure?" Be very sure to review what you are about to do *before* pressing Enter to run the command.

In this section, we'll assume that the partition you want to grow is the last one on the disk (this will usually be the /home partition, where you store your media) and is the last one you created when installing (and has the highest partition number). You can look at the partition layout using the Gnome Partition Editor, which is included on the Ubuntu installation CD, or you can install it manually by installing the gparted package with aptitude or the graphical Synaptic.

To move your data to the new hard disk, shut down your MythTV box, and boot off the Ubuntu installation CD or DVD. Copying a disk that you're currently using does not work, which is why you are rebooting off the Ubuntu installation disk in this example.

You will need to find out the device names of the hard disks you have connected to your machine (which is the source and destination disk). You want to look in the Gnome Partition Editor (you'll find it under Administration in the System menu) to see which disk is blank and which is the disk with the partitions with your beloved data. The menu on the top right of the window shows the device names. In the example shown in Figure 10-1, /dev/sda is the source disk, and /dev/sdb is the destination (blank) disk. Commonly, IDE hard disks are /dev/hda (or hdb, hdc, hdd, and so on), and SCSI (or SATA) disks are /dev/sda (or sdb, sdc, and so on).

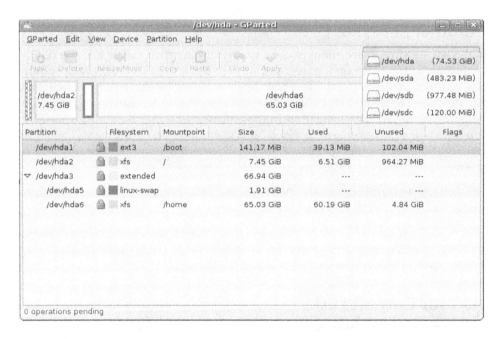

Figure 10-1. *Gnome Partition Editor (GParted), showing the list of available drives*

You now want to open a terminal window from Applications ➤ Accessories ➤ Terminal. The next step is to start the copy. We'll show how to copy the complete contents of the original hard disk to the new one. In this example, the *source* is /dev/sda (where we are copying *from*), and the *destination* is /dev/sdb (where we are copying *to*).

Caution We were serious about it being possible to lose data during this operation. Make sure you have a recent backup and that you are copying from the right device to the right device.

Start the copy with this:

```
$ sudo dd if=/dev/sda of=/dev/sdb bs=1M & PID=$!
$ sudo kill -USR1 $PID
```

The first command starts /bin/dd to copy the data, puts it in the background (the & character does this), and then saves its process ID in the shell variable PID. The second command makes the dd process print a status report like this:

```
9938736+0 records out
5088632832 bytes (5.1 GB) copied, 122.7450 seconds, 41.5 MB/sdy
```

You can run that second command more than once if you want another update on the progress of the copy. You'll know the dd command is finished when the output of the jobs command is empty. Depending on the speed and size of your source disk (usually the destination will be newer and therefore faster), this could take a few minutes to several hours. The status information printed after running the previous kill command will tell you at what rate the copy is going. You can work out from this rate approximately how much longer you will have to wait. The 1M means that dd should read and write in 1MB blocks (1,048,576 bytes); you can increase this if you find that your throughput benefits from this.

You now need to change the size of the partition. You can use the text-based cfdisk program to do this (see Figure 10-2 for an example of how this program looks). Launch it like this (where /dev/sdb is the device name of the new disk):

```
$ sudo cfdisk /dev/sdb
```

Figure 10-2. cfdisk *with the last partition selected*

Use the left and right arrow keys to select the function, and return to execute that function. Use the up and down arrow keys to select a partition. Select the last partition on the disk, and select Maximize to expand that partition to take all the remaining disk. When you're done, write the partition table to disk and quit.

You are now ready to resize the file system. This will allow you to use the extra space. For XFS disks, this is simple. You first mount the partition:

```
$ sudo mount /dev/sdb2 /mnt
```

and then run xfs_growfs:

```
$ sudo xfs_growfs /mnt
```

The last line of the xfs_growfs output will show you the old and new numbers of data blocks. You can now shut down the system, swap the disks over (that is, remove the old one and put the new one in the same place), and restart your MythTV box. You will now find you have more disk space available. If you use the ext3 file system, you will have to unmount the partition before resizing. To unmount the partition, you cannot have any open files on it. A good way to do this is to reboot without logging in, switch to a console using Alt+Ctrl+F1, log in there, and change directories. The following is how to resize /dev/hda2, which is mounted at /home. Check the output of the mount command to find out what device you have mounted.

```
$ cd /
$ sudo umount /home
$ sudo resize2fs /dev/hda2
$ sudo mount /home
$ logout
```

The output of the resize2fs command will tell you how large the file system now is. You can now switch back to the graphical login screen (use Ctrl+Alt+F7) and log in. You'll notice that you now have more disk space.

Copying One Partition at a Time

If your main data partition is not the last one on the disk, you should create new partitions on the new disk and copy each partition one by one (then using xfs_growfs, or similar) for each partition. This means running dd for each partition:

```
$ sudo dd if=/dev/sda1 of=/dev/sdb1 bs=1M& PID=$!
$ sudo kill -USR1 $PID
```

You will also need to reinstall the bootloader (the bit of software that sits at the start of the disk and loads Linux).

To reinstall the bootloader, swap the new disk in, take out the old, and then start your system from the Ubuntu installation CD. There, open a terminal. Mount the root partition of your MythTV box:

```
$ sudo mount /dev/sda2 /mnt
```

Now, mount the /proc file system, as well as bind the /dev directory so you get access to the devices:

```
$ sudo mount -o bind /dev /mnt/dev
$ sudo mount -t proc none /mnt/proc
```

Then use chroot to change the root directory of the file system to the new disk. This gives you a shell that is (almost) like you started a system from the disk to which you chrooted:

```
$ sudo chroot /mnt
```

Now, install grub:

```
# grub-install /dev/sda
```

Now, if you reboot, you should start up into your MythTV system with access to more disk space. If you are inexperienced with Linux, then you are probably best off asking your local Linux user group to help you out with process, because it is dangerous to your data.

Adding Another Disk with LVM

If you are using LVM, adding more disk space is easy. You first need to create a physical volume on the new disk. You will need to replace the device name, volume group name, logical volume name, and mount point in these examples with those from your own system; you can retrieve most of this information from the output of df -h. In the following example, you can see that the MyVolumeGroup volume group has a logical volume named home that is mounted on /home. You can also see other logical volumes and their mount points: root on /, tmp on /tmp, var on /var, and usr on /usr.

```
$ df -h
Filesystem           Size  Used Avail Use% Mounted on
/dev/mapper/MyVolumeGroup-root
                     1.5G  215M  1.3G  15% /
varrun               499M   96K  498M   1% /var/run
varlock              499M     0  499M   0% /var/lock
udev                 499M  116K  498M   1% /dev
devshm               499M     0  499M   0% /dev/shm
/dev/md0             243M   35M  197M  15% /boot
/dev/mapper/MyVolumeGroup-home
                      73G   71G  1.9G  98% /home
/dev/mapper/MyVolumeGroup-tmp
                    1014M  3.5M 1011M   1% /tmp
/dev/mapper/MyVolumeGroup-usr
                     3.0G  1.7G  1.4G  56% /usr
/dev/mapper/MyVolumeGroup-var
                     3.0G  1.5G  1.6G  49% /var
```

So, you can now make the /dev/sdb volume usable as a physical volume:

```
$ sudo pvcreate /dev/sdb
```

Then you need to add it to the volume group:

```
$ sudo vgadd /dev/sdb MyVolumeGroup
```

Then you can expand the logical volume to use more of the space:

```
$ sudo lvextend -L +80GB MyVolumeGroup/home
```

Now you can expand the file system. With XFS, this is an online operation (you do not need to unmount the file system; in fact, you must have it mounted):

```
$ sudo xfs_growfs /home
```

For ext3, you need to unmount the partition and use resize2fs. For XFS, this should complete almost instantly and print a report of the current state of the disk. We discussed the exact details of using these commands in the "Replacing Your Current Disk" section earlier in the chapter. You'll notice the increase in data blocks on the file system. If you run df -h, you will now see you have more free space on that file system.

```
$ df -h
Filesystem            Size  Used Avail Use% Mounted on
/dev/mapper/MyVolumeGroup-root
                      1.5G  215M  1.3G  15% /
varrun                499M   96K  498M   1% /var/run
varlock               499M     0  499M   0% /var/lock
udev                  499M  116K  498M   1% /dev
devshm                499M     0  499M   0% /dev/shm
/dev/md0              243M   35M  197M  15% /boot
/dev/mapper/MyVolumeGroup-home
                      153G   71G 81.9G  46% /home
/dev/mapper/MyVolumeGroup-tmp
                     1014M  3.5M 1011M   1% /tmp
/dev/mapper/MyVolumeGroup-usr
                      3.0G  1.7G  1.4G  56% /usr
/dev/mapper/MyVolumeGroup-var
                      3.0G  1.5G  1.6G  49% /var
```

Using Network-Attached Storage

You might decide that having storage on another machine (or perhaps attached to a little NAS device such as the Linksys NSLU2) is the setup for you. In this case, you can copy your existing data to the network volume, remove it from the local drive, and mount the network volume as the data directory. Using network-attached storage in addition to local storage will be a lot easier with the Storage Groups feature of MythTV 0.21.

Adding Remote Backends

Adding a remote backend to your MythTV setup allows you to perform several tasks. You can use these backends to process commercial flagging and transcoding jobs (for example, purely as extra CPU power). You can install TV capture cards in them and record shows onto their local disks to be able to record more shows at once.

You first need to install MythTV on the new backend. Unlike your initial setup, what you enter in `mythtv-setup` will be slightly different. The first step, selecting what language, is the same (see Figure 10-3). Before running `mythtv-setup`, however, you should make sure the new backend can connect to the master MySQL database on your main MythTV box by granting the right permissions:

```
$ mysql -u root
mysql> grant all on mythconverg.* to mythtv@% identified by 'mythtv';
```

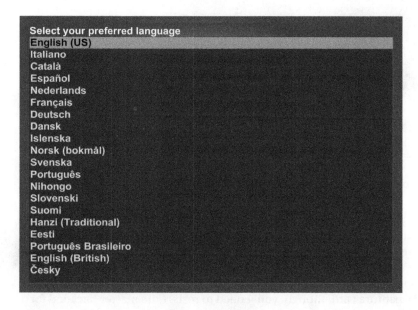

Figure 10-3. *Selecting the language in* mythtv-setup

To grant permission to the database for only one machine, replace % in the previous query with the IP address of the machine. Leaving % there indicates that the mythtv user can connect to the database from any IP address. The word after `identified by` is the password; you might want a more secure one than simply `mythtv`.

The second configuration screen enables you to connect to the database (see Figure 10-4). Enter the correct user, host, and password. If you have connection failures, check that the password is correct and that you don't have `skip-networking` or `bind-address` set in `/etc/mysql/my.cnf` on the master backend that runs the MySQL database. You can find more information about how to configure MySQL for network access available in Chapter 8.

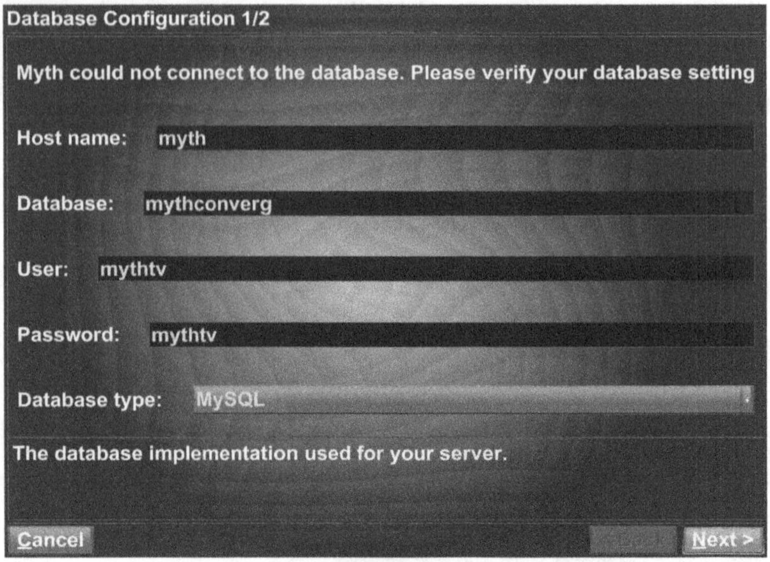

Figure 10-4. *Connecting to the MySQL database server*

Now that you've managed to connect to the database, you are able to run mythbackend, and it should connect to the master backend and be able to process jobs such as commercial detection and transcoding. You can check the logs on the master backend to see what your slave backend has been doing. If you are going to have capture cards on this backend, follow the same instructions as adding capture cards for adding capture cards for your master back-end. After you've added a capture card, though, you'll need to restart the master backend for the changes to take effect.

Conclusion

Expanding your MythTV setup is something that you'll probably want to do once you've gotten used to being able to record all your favorite shows. You can expand your installation in a few ways, which we've covered in this chapter, from adding more tuners and backend machines to adding more disk space to store programs. In the next chapter, we will discuss how to play and create your own DVDs, so keep reading.

CHAPTER 11

■■■

Using MythWeb: A Web Interface to MythTV

MythWeb is a web-based interface to MythTV. It lets you (among other things) schedule recordings, check upcoming recordings, view program listings, check backend status, and change some configuration parameters. It's commonly used to manipulate the recording schedule from computers not running a MythTV frontend, including over the Internet when away from home; though, if you have a single-frontend system, it can be helpful for setting up recordings while you're watching a program.

MythWeb is a web application that runs on your backend. It requires some software on the backend to run (PHP and Apache), which we'll show how to install. It also requires some basic configuration so it knows how to connect to your MythTV setup. Using it, however, is rather simple and can be done from a standard web browser.

Installing MythWeb

Although MythWeb is part of the MythTV plug-ins, it's not installed in the same way. The first step is to install the software required to run the web application: the web server, the language that MythWeb is written in (PHP), and the module that lets it talk to the MySQL database. This is simple on Ubuntu with the following command:

```
$ sudo apt-get install apache2 php5 php5-mysql
```

You need to make a small configuration change to the Apache web server so that the configuration options for MythWeb can be loaded correctly. Specifically, you need to edit the /etc/apache2/sites-enabled/default file with a text editor. (The location and name of this file and the other configuration files for Apache and PHP will vary widely if you're using a distribution other than Ubuntu.) For example, to edit it using the Gnome text editor, run this:

```
$ sudo gedit /etc/apache2/sites-enabled/default
```

You will see a section in the configuration file like this:

```
<Directory /var/www/>
        Options Indexes FollowSymLinks MultiViews
        AllowOverride None
        Order allow,deny
        allow from all
        # Uncomment this directive if you want to see apache2's
        # default start page (in /apache2-default) when you go to /
        #RedirectMatch ^/$ /apache2-default/
</Directory>
```

A few lines before this section in the configuration file is the `DocumentRoot` configuration option. This option tells Apache where to look for files to serve. If you accessed `http://server/file.html`, Apache would look in `DocumentRoot` for `file.html`. For a `DocumentRoot` of `/var/www`, Apache would look for `/var/www/file.html`.

This section configures some options for the `/var/www` directory (which is being used as the `DocumentRoot` option). Currently, the `Indexes`, `FollowSymLinks`, and `MultiViews` options are enabled. The `AllowOverride` option being set to `None` means that a configuration file in the `/var/www` directory cannot change any options. For MythWeb to function, you should change this to `Allow Override All`. If you do not enable this, you might get a strange error from Myth-Web such as "Please install the MySQL libraries for PHP."

It is also possible you will want to (either now or later) configure a `RedirectMatch` line (like the one commented out in the earlier code) to make going to `http://server/` automatically redirect to `http://server/mythweb/`. Obviously you wouldn't do this if you use the web server for anything else. Such an option would look like this:

```
RedirectMatch ^/$ /mythweb/
```

You also need to uncomment a line in the `.htaccess` file of MythWeb (which, after you copy the MythWeb files across, will be `/var/www/mythweb/.htaccess`); it's called `RewriteBase`. You will also need to enable the `mod_rewrite` Apache module. You can do this by running the following command:

```
$ sudo ln -s /etc/apache2/mods-available/rewrite.load /etc/apache2/mods-enabled/
```

If you don't have the `mod_rewrite` module enabled, you will get an "Internal Server Error" when trying to access MythWeb. You will also get a more detailed error message in the Apache error log (`/var/log/apache2/error.log`). Obviously, you won't want to do any of this if you plan to use the Apache server for other purposes, such as running PHPMyAdmin.

To enable the MySQL support for PHP 5 (which you installed earlier), you need to configure the `php5-mysql` package. Since you've already installed the package, the technical term is that you are going to *reconfigure* the package. It will ask you two questions: Do you want to enable MySQL in PHP 5 for Apache? Do you want to enable it for the command-line interface (CLI) PHP 5? Answering "yes" to both is a good idea. To reconfigure `php5-mysql`, run the following command:

```
$ sudo dpkg-reconfigure php5-mysql
```

You now need to restart Apache to make the configuration changes take effect. Do this by executing this command:

```
$ sudo /etc/init.d/apache2 restart
```

You are now ready to copy the MythWeb files to /var/www/ and test your Apache configuration. From the mythplugins-0.20a directory, execute the following:

```
$ sudo cp -r mythweb /var/www/
```

If you have made any changes to the default install, such as where the MySQL server is running and what user MythTV connects to the database as (or what password it uses), you will need to change these in the /var/www/mythweb/.htaccess file. If you enabled RewriteRule earlier, you will need to enable RewriteBase in this file as well. Otherwise, you can continue.

The last step you have to take is to change the owner of the /var/www/mythweb/data directory so that MythWeb can store information in there. You do this by executing the following command:

```
$ sudo chown www-data /var/www/mythweb/data/
```

If you don't do this, MythWeb will report an error message that it cannot write to the data directory. You can now use a web browser to open http://*server*/mythweb/ and see the MythWeb home page. You should replace *server* with the IP address of your MythTV box (or host name if you have it configured correctly). You will see a page similar to Figure 11-1.

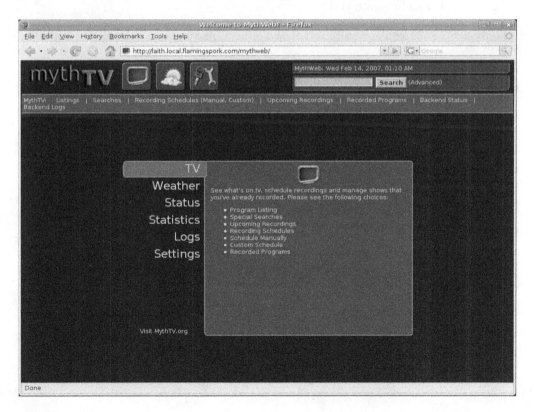

Figure 11-1. *The MythWeb home page viewed from Firefox on Ubuntu*

If you do not see a page similar to Figure 11-1, go back and check every step of the configuration, starting with what error message you get. Before exposing your MythWeb installation to the Internet, you need to read the instructions in the "Securing MythWeb" section later in this chapter so that your installation is password protected and so random people and search engines cannot delete your recordings!

Using MythWeb

You can perform many tasks from MythWeb, including viewing TV listings, viewing and changing upcoming recordings, checking the backend status, and looking at some statistics of your recording habits. At the top of each MythWeb page are three icons (Figure 11-2) for the three parts of MythWeb: TV functions, weather, and MythWeb and MythTV settings.

Figure 11-2. *The MythWeb icons for TV functions, weather, and settings*

Across the top of each page is a textual listing of the main pages of the TV functionality of MythWeb (shown earlier in Figure 11-1):

Listings: This shows the TV listings and from there allows you to add or change recording schedules.

Searches: This lets you search the TV listings. The authors tend to use the search box at the top right of the page instead of these preset searches.

Recording Schedules: This lists all the recording schedules you have set up and allows you to change them.

Upcoming Recordings: This is similar to the screen of the same name in the MythTV frontend but on a website. This was formerly called Scheduled Recordings but was changed because that was somewhat confusing.

Recorded Programs: This lists all the recorded programs and allows you easily to delete programs as well as download the video files for playback in a compatible player.

Backend Status: This gives you an overview of the status of your backends. It lists what the upcoming recordings are, what jobs are queued, and how much free disk space there is on each backend (and in total).

Backend Logs: This shows detailed logging of what the backends have been doing.

Using MythWeb TV Listings

Selecting Listings either from the text menu or from the MythWeb home page takes you to a view similar to the program guide in the MythTV frontend (see Figure 11-3). You can quickly jump to any date and time using the drop-down menus on the top right. By default, the current time is shown.

By placing your cursor over the title of the showing, a box pops up showing details of that program, including why the recording might not be being recorded. A border around a TV show indicates there is a recording schedule for that show. A bold solid line indicates that the showing will be or is being recorded (see Figure 11-4). A red border indicates a recording conflict and that this showing won't be recorded (see Figure 11-4). In Figure 11-4, the *Firestarter* program isn't being recorded because another program with a higher priority is being recorded at the same time. In Figure 11-3, the showing of *Alias* is being recorded but with a note that this specific showing was manually set to be recorded. A dotted border indicates that this showing will not be recorded because there's another showing that is (see Figure 11-5).

Figure 11-3. *The MythWeb TV listing*

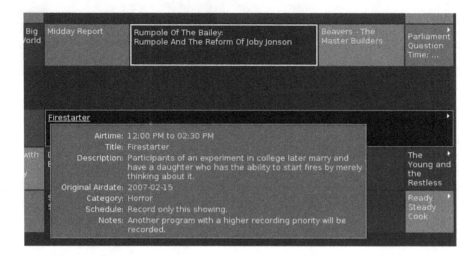

Figure 11-4. *The MythWeb TV listing with a simultaneous recordings (only one will be recorded)*

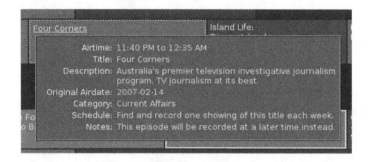

Figure 11-5. *A showing that isn't being recorded because there is a later showing*

When you click the title of a showing, MythWeb takes you to the program detail page for that program. From here you can see the description and possible conflicts as well as change the scheduling options.

Using the MythWeb Program Details

Figure 15-6 shows the program details page for a recording. This lets you view your current settings as well as change most of the MythTV scheduling options for the program. You can also use this page to set up a new recording schedule. This is especially useful when you are away from home and somebody tells you about this great new show; you can log onto MythWeb remotely and set your MythTV box to record that program.

From here it's easy to see possible conflicts (the box on the lower left). In this case, you can see that *Little Britain* won't be recorded because of a higher-priority program being recorded (on the right you can see that *Veronica Mars* has a recording priority of 6). You can also see that *Lateline* won't be recorded, and if you put the cursor over the title, you'll see the reason (an earlier recording will be recorded instead).

The three links on the bottom left let you quickly jump to other useful places. The What Else Is on at This Time? link will take you to the TV listing page at the appropriate time. Find Other Showings of This Program will perform a search of the guide data for showings titled *Veronica Mars*, and Back to the Program Listing does exactly what it says and takes you back to the program listing. Because of a bug in the version of MythWeb that shipped with MythTV 0.20, it's possible that this link won't take you back to the listing page for the same time period as the show.

All the schedule options detailed in the box on the right are the same as you would set with the MythTV frontend. There are a few small oddities, though, that are different from what is displayed in the frontend. If No. of Recordings to Keep is set to zero, it means an unlimited amount. The start early and late times being zero means the default amount of time. In Figure 11-6 you can see we've set the recording to end 37 minutes late (why we choose 37 is a mystery) because it's screened later at night, which usually means (for Australia) that the TV channel will be running late. One other difference is the ease of setting a schedule override, which in MythWeb simply involves selecting Record This Specific Showing or Do Not Record This Specific Showing, and it's rather obvious that it's an override.

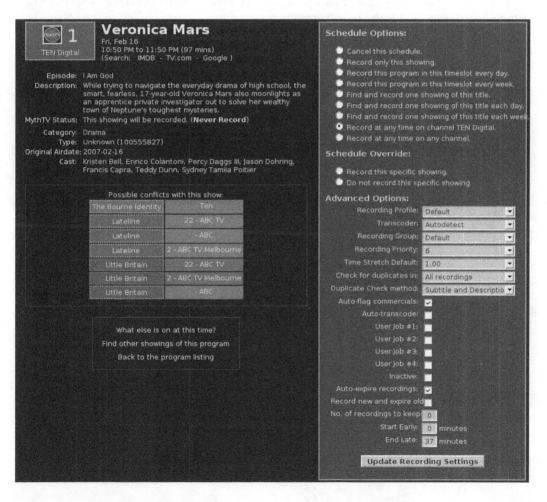

Figure 11-6. *MythWeb program details*

Using Searches

You can start a search for a program several ways: by entering a search term in the search box on the top right or by clicking many of the links throughout the MythWeb interface (for example, the Find Other Showings of This Program link from the program details page). By default, the search is on all the guide data, so typing **Homer** will find any program that has *Homer* in any of the fields, including the description. A search for a term such as *crime* will bring up listings in the Crime Drama category, documentaries that have *crime* in their descriptions, and the odd movie that has something to do with crime. Searching works well when you receive a lot of channels or can't really remember what some programs are called.

You can also place other constraints on the search such as matching only high-definition recordings; matching only those shows on commercial-free channels; putting AND and OR match constraints on the title, subtitle, description, category, or channel; specifying when the showing is aired; specifying when the program originally aired; and specifying the type of program. The ability to perform good searches can depend on the quality of your guide data, though.

Figure 11-7 shows part of the results for a search on *crime* from the search box. This type of simple search is probably the most common type that people use.

Figure 11-7. *MythWeb searching for crime*

Using MythWeb Recording Schedules

Selecting Recording Schedules from the text menu at the top of any of the MythWeb pages takes you to a list of all the recording schedules you have set up in your MythTV backend. This includes all the schedules for programs that aren't currently in the TV listing as well, so you might see listings for programs that aren't currently airing (for example, we see a schedule for the 2006 FIFA World Cup). To sort the list of recordings, you can click the titles of the columns in the table (title, recpriority, channel, profile, transcode, recgroup, and type). Clicking the title again will sort in the reverse order. In Figure 11-8 we've sorted the list by Recording Priority (recpriority) with the highest at the top. We also have the mouse over the *Veronica Mars* recording schedule to see some information about it.

By clicking the name of a program, you are taken to a details page about that recording schedule. This page is almost the same as that discussed in "Using the MythWeb Program Details" section earlier in this chapter.

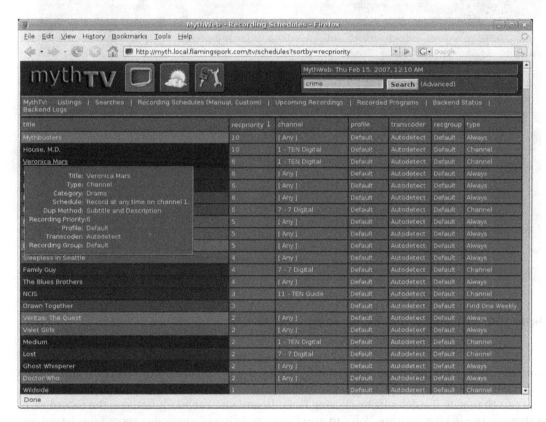

Figure 11-8. *MythWeb recording schedules*

You can also use the Manual or Custom Recording Schedule pages (in brackets next to the Recording Schedules link in the text menu) to create either a manual schedule at a specific time or a custom schedule based on some search criteria in a way similar as in the MythTV frontend.

Checking the MythWeb Upcoming Recordings

One of the most useful features of MythWeb is being able check the list of upcoming recordings and make any modifications necessary (such as adding overrides). Selecting Upcoming Recordings from the text menu takes you to a page with a table of the upcoming recordings in the order of their starting times (see Figure 11-9). Again, clicking any of the titles of the table will reorder the table in that order.

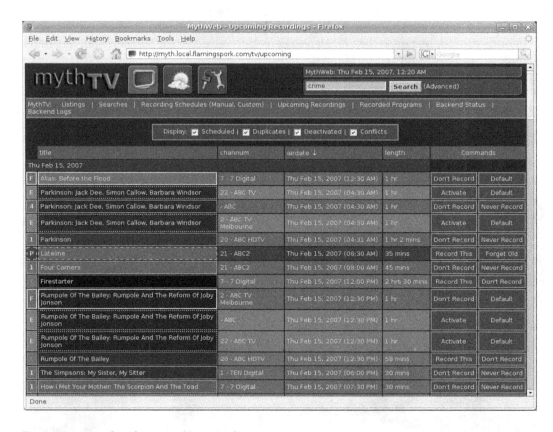

Figure 11-9. *MythWeb upcoming recordings*

Selecting a program name will take you to the same program details screen discussed previously. The additions that are really useful here are the command buttons on the right of each recording. You can easily add an override to disable a specific showing by clicking Don't Record. You can change to the defaults by clicking Default, and you can easily activate the recording of a showing by clicking Activate. You can also filter the listing of upcoming recordings based on a few options (the checkboxes at the top of the list).

Checking Your Recorded Programs

The Recorded Programs page lists all the programs you have recorded and stored on a backend. This page can take a little while to display because the preview thumbnails for recordings might need to be generated. Figure 11-10 shows a list of recorded programs. You can again sort this list by clicking the title of the table columns. At the bottom of the page is a brief summary, such as "92 programs, using 158 GB (4 days 7 hrs 31 mins) out of 226 GB (68 GB free)."

Figure 11-10. *MythWeb recorded programs*

For each recording, the title, subtitle, description, channel, airdate, and length appear. Also, the file size on disk appears. You get no indication of whether this recording has gone through transcoding apart from the file size. You can click the Delete or Delete + Rerecord button to remove the recording. It is often much easier to delete many programs from MythTV through the web interface than with a remote control on the frontend. You can also change whether the recording should autoexpire via a checkbox. The other information shown includes whether the recording has had commercial flagging performed. However, if a recording has had the commercial detection turned into cutpoints and has since been transcoded, "have_commflag" will be set to No. This is the same with "has cutlist"; a recording that has been transcoded no longer has a cutlist.

One very different feature of the Recorded Programs page is that if you click the preview icon, the title, or the subtitle, you'll start to download the recording instead of being taken to the program details page. These recordings can be streamed by some media players (such as Xine) or saved to disk and played back. Although potentially useful, both the authors run a MythTV frontend on computers instead of using this function because of the extra functionality it provides (some media players won't be able to skip through the stream).

Checking Your Backend Status

The Backend Status page is useful to monitor what your backends are currently doing (and whether they are online). Figure 11-11 shows the Backend Status page where you can see that Encoder 1 is recording *Alias* and Encoder 4 is online but not recording. The Schedule listing shows upcoming recordings. If you put your cursor over the entry in the table, some details will appear about the recording schedule. You might note that the encoders on this machine are numbered 1 and 4. MythTV numbers them according to the order in which they're added to the machine; if you later remove one or you make a mistake during setup and have to delete one, that number will be left empty.

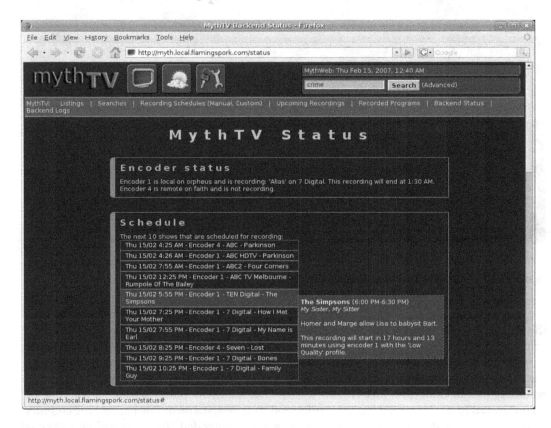

Figure 11-11. *MythTV backend status*

Further down on the Backend Status page you can see the current job queue (see Figure 11-12). Again, placing your cursor over the entry shows details. For example, here you can see that a transcode job completed for a recording of *Malcolm in the Middle* that reduced the storage size from 2.4GB to only 298.9MB. The red color of the word *Transcode* on the first entry of *The Simpsons* indicates an error occurred, while the green *Transcode* next to the first *My Name is Earl* episode indicates that this job is currently being run. Plain white means the job completed successfully, as you can see.

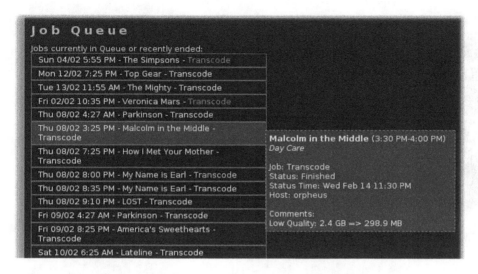

Figure 11-12. *MythWeb backend status job queue*

The final part of the Backend Status page is the Machine Information section (see Figure 11-13). Here you can view the load average of the machine that MythWeb runs on as well as the free disk space on the master backend and any secondary backends (here the backend running on the machine called *faith*). If possible, MythWeb also attempts to retrieve the current temperature of the CPU, but the accuracy of this might vary from machine to machine. Also useful is the information on when `mythfilldatabase` last ran and for how long there is guide data.

The load average is a standard Unix metric for the load on a machine; technically, it's the number of processes in the Run queue, averaged over 1, 5, and 15 minutes. If the load average on your machine gets too high, it will start to bog down, perhaps skipping during playback, or even causing recordings with bad spots in them. The effect will be worse on machines whose tuner cards do not do hardware-MPEG encoding. Over time, you'll acquire a feeling for how high a load average is safe on your machine; one example is an AMD Athlon 2500+/Via KT600 machine with 512MB of RAM and 3 Hauppauge tuners with IVTV 0.4.0 drivers, which can tolerate a load average from 5 to 6 without breaking recordings. If your 1-minute load average is higher than your 5-minute load average, then the average is going up, and you might have a problem you need to fix; if it's lower, then it is probably descending. Runaway X servers and web browsers are common reasons for problems. Occasionally a locked up frontend can cause this problem as well.

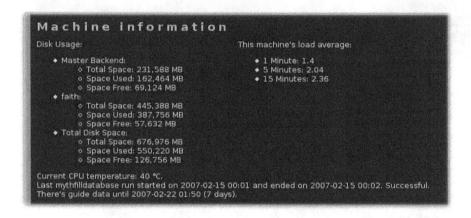

Figure 11-13. *MythWeb backend status machine information*

Checking Your Backend Logs

The Backend Logs page simply shows a reverse chronology of log events in the backend logs. No sorting is supported, but looking here can be useful in the event of an error or to check that things really are happening.

Securing MythWeb

The easiest way to add basic password authentication to MythWeb is to use HTTP authentication. This isn't the most secure form of authentication, but it's rather easy to set up. First, edit the .htaccess file in the mythweb directory (most likely /var/www/mythweb/.htaccess). For example:

```
$ sudo gedit /var/www/mythweb/.htaccess
```

Then, add the following configuration options (possibly near the example in the .htaccess file):

```
AuthType        Basic
AuthName        "MythTV"
AuthUserFile    /var/www/mythweb/.htpasswd
Require         valid-user
```

You will now need to create the .htpasswd file that will store the usernames and encrypted passwords for people who can access the MythWeb installation. We will use the htpasswd program for this example. The first time it's run, you'll need to use the -c (create) option to create the file. In subsequent runs, if you want to add users, you do not need to specify -c, and if you do, it will overwrite the file. For example, to add a user called *myth* (you will be prompted for the password), run the following:

```
$ sudo htpasswd -c /var/www/mythweb/.htpasswd myth
```

To add another user called *foobar*, you would run this command:

```
$ sudo htpasswd /var/www/mythweb/.htpasswd foobar
```

Now, when you next browse to the MythWeb installation, you will need to enter a valid username and password (as shown in Figure 11-14). If you get an Internal Server Error page, check the Apache error log (`/var/log/apache2/error.log`) for clues—you might have made a typo in the configuration file.

Figure 11-14. *Firefox asking for a password to access MythWeb*

To be clear, you don't need to set up authentication to protect only against malevolent (or bored) humans; MythWeb makes deleting a recording a function controlled by retrieving a link. Therefore, if a search engine spider gets ahold of your MythWeb server address, it can delete everything on your machine, with no human intervention. This has happened to several people. If you're going to make your MythWeb server accessible to the Internet (and, really, why would you not?), secure it.

Conclusion

In this chapter, we covered most of the usage of MythWeb, including the not so simple installation procedure as well as how to set up password authentication if you are going to make your MythWeb installation viewable over the Internet (to provide some level of security). In future versions of MythTV, it is likely that the functionality of MythWeb will be expanded to be even more useful. In its current form, it is extremely useful, and many people have found that running it greatly improves their MythTV experience, especially if you travel a lot and want something to watch when you get home.

CHAPTER 12

■■■

Working with DVDs

This chapter discusses two aspects of DVDs and MythTV. First off, we'll look at how to play and import DVDs with MythTV, and then we'll look at how you can create your own DVDs of things you have recorded with MythTV. *Importing* (sometimes referred to as *ripping*) is the process of taking the data on a DVD and putting it on the machine's hard disk, so that you don't have to have the DVD in the drive to play it. This can be particularly convenient for DVDs that you like watching a lot, or that are slightly damaged. You can also optionally transcode these ripped files to a format that will take up less space on the disk.

There are two ways to burn DVDs with MythTV: MythBurn and MythArchive. We only discuss MythArchive, as it is better supported with modern versions of MythTV (MythBurn hasn't had a new release in a while now); also, MythArchive is now included as part of MythDVD in MythTV 0.20.

Playing DVDs and VCDs

DVD and VCD playback is handled in MythTV with the MythDVD plug-in. This plug-in was built and installed in Chapter 9. This plug-in provides a simple user interface for starting playback of a DVD or VCD, and archiving DVDs to disk. Once you've started DVD or VCD playback, the actual playback is handled by either the internal player or an external program, which can be pretty much anything capable of being started from a command line. Common examples of DVD playback applications used with MythDVD are xine and mplayer. If you read Chapter 9, then you'll note that this is very similar to the way that the MythVideo plug-in works, with the notable exception that you can only have one DVD playback command line specified.

WHAT IS A VCD?

Before DVDs were as cheap as they are now, some people would archive their video to a format called VCD (Video CD). This format uses a CD-ROM disk to store the data, which made it much cheaper than DVD was at the time. Now that DVD media is much more common, VCD isn't used as much any more.

All you really need to know about VCD is that it's like DVD, but on a CD-ROM, and that the video quality isn't as high as DVD.

You configure MythDVD from Utilities/Setup ➤ Setup ➤ Media Settings ➤ DVD Settings ➤ General Settings (see Figure 12-1 for an example), where you specify the DVD device file (in my case, /dev/hda—in many cases, there will be a symlink at /dev/dvd that you can use—that doesn't work for me because I have more than one DVD drive, and don't want to use the default as the DVD device with MythTV), the VCD device file (in my case, my CD-ROM drive at /dev/cdrom), and what to do when a DVD is inserted into the computer.

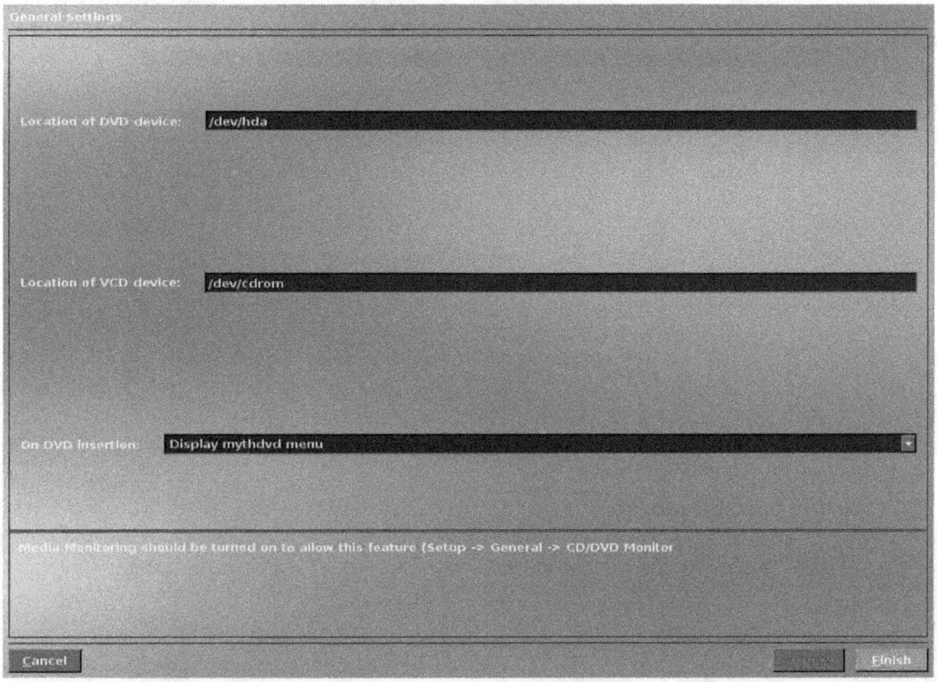

Figure 12-1. *General configuration options for MythDVD*

The options for what to do when a DVD is inserted are as follows:

- Do nothing.

- Play the DVD.

- Import the DVD.

- Go to the MythDVD menu.

I personally prefer to just go to the MythDVD menu, because the action I want to take with a newly inserted DVD changes based on my needs. Next, you configure the playback settings at Utilities / Setup ➤ Setup ➤ Media Settings ➤ DVD Settings ➤ Play Settings (see Figure 12-2).

Here you configure the DVD player program to use, and on the next screen you configure which player to use for VCDs (see Figure 12-3).

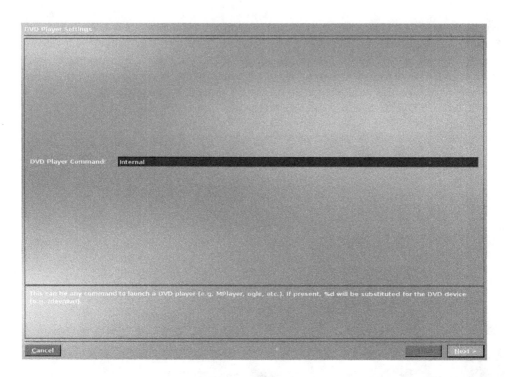

Figure 12-2. *Configuring the DVD player*

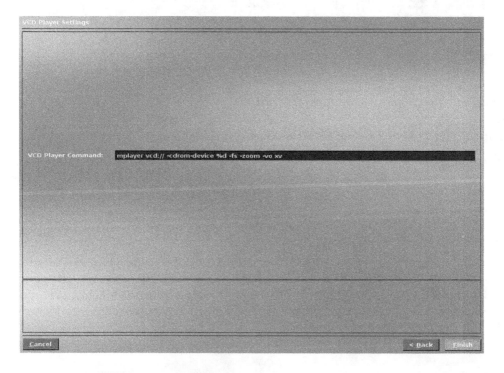

Figure 12-3. *Configuring the VCD player*

There are a variety of player options for DVDs and VCDs, all of which are covered in the following "Selecting Player Applications" section. I recommend you stick with the internal player for now, though. Finally, you're ready to actually play a DVD. You do this by going to the Optical Disks ➤ Play DVD menu, which will launch your selected DVD player application.

Selecting Player Applications

You are spoiled for choice with DVD player applications, and which one you use will end up being mostly dependent on personal preference. We will focus on the three most common choices, although you can use any application that plays DVDs and can take command-line arguments to tell it what to do. I personally used xine with MythTV 0.19, which I recommend because of its support for lirc, DVD menus, and excellent playback. With the introduction of MythTV 0.20, I now use the internal player—it supports DVD menus and is extremely well integrated into the rest of MythTV.

The Internal Player

The biggest advantages of the internal DVD player shipped with MythTV are that it's simple to set up and integrated with the rest of the MythTV user interface. In fact, the player looks like the rest of MythTV, and apart from your having to go to a different menu to play the DVD, it's just like watching a TV recording. Importantly, this means that the back key, volume buttons, and playback controls such as pausing and skipping all work as you would expect them to. Additionally, the player loads very quickly because it's not a separate application. Finally, you don't need to do any additional lirc setup to get DVD remote control working.

However, there used to be significant disadvantages to the internal player. It didn't support DVD menus, which made it hard to navigate between episodes if you were watching a DVD containing television shows. That's less annoying with movies, though. Additionally, some of the skipping behavior doesn't map to what you would expect from a normal DVD player. However, the DVD menu issue has been addressed in MythTV 0.20, which is new at the time of writing this chapter. This means that if you're running MythTV 0.20, then the internal player might be a reasonable choice.

To use the internal player, just leave the playback command line at Utilities / Setup ➤ Setup ➤ Media Settings ➤ DVD Settings ➤ Play Settings blank.

mplayer

mplayer is very similar to the internal DVD player that ships with MythDVD. It is fast to load, plays almost any file with ease, and works well. However, it doesn't support DVD menus, which makes it hard to use in some cases. To me, not having menu support is a bit of a showstopper.

To install mplayer, simply run this command:

```
$ sudo apt-get install mplayer
```

To use mplayer, put the following text in the playback command line at Utilities / Setup ➤ Setup ➤ Media Settings ➤ DVD Settings ➤ Play Settings:

```
mplayer dvd://1 -dvd-device %d
```

This command line will launch mplayer and start playing title 1 on the disk, using the device specified in the DVD configuration as mentioned earlier.

xine

xine is what I used to use for my DVD playback with MythTV before the internal player improved with MythTV 0.20. It supports DVD menus, plays every DVD I have ever asked it to play (including a couple of DVDs that cheap hardware DVD players had trouble with), supports decoding DVDs from different regions (I have a lot of Australian DVDs that I still want to play now that I live in the United States), and has good remote control support. To get the remote control working, just follow the instructions for setting up lirc in Chapter 3. The biggest disadvantage of xine is that it takes a few seconds to load when you start playing a DVD.

To use xine, put the following text in the playback command line at Utilities / Setup ➤ Setup ➤ Media Settings ➤ DVD Settings ➤ Play Settings:

```
xine -pfhq --no-splash dvd:/%d
```

Playing Region-Encoded DVDs

One of the more frustrating aspects of DVD ownership is that the disks are sometimes locked to a specific region by the manufacturer (this is despite the fact that locking/unlocking behavior is illegal in some areas). If unlocking such DVDs is legal for you, then it's a relatively simple process. The secret is to decode the encryption on the DVD with software, instead of the hardware inside your DVD player (which will be locked to a region). To do this, you need the CSS decryption software, which isn't shipped with the default Ubuntu installation, for legal reasons. You can download and install it like this:

```
$ apt-get install libdvdread3
$ sudo /usr/share/doc/libdvdread3/examples/install-css.sh
```

This will leave you with the capability to play region-encoded DVDs in software players like xine.

A BRIEF HISTORY OF SOFTWARE DVD DECRYPTION

Jon Lech Johansen and a team of others were the original authors of the DVD decryption software DeCSS, which is now commonly used to watch legally purchased DVDs on a variety of operating systems. Jon's work was motivated by his desire to watch DVDs he had purchased on Linux, and resulted in him being prosecuted under Norwegian law. After a conviction and victory at appeal, Jon has been exonerated of these charges.

You can find more information about Jon on Wikipedia (http://en.wikipedia.org/wiki/DVD_Jon) and his blog, at http://nanocrew.net/.

Importing DVDs

Importing DVDs gives you a convenient way of watching a disk without having to put it into the drive. This can be especially useful for disks that you watch a lot—for example, if you have children who have favorite shows. The DVD import process works with the assistance of a transcoding daemon called mtd, unless you are only willing to use "perfect" copies, which are much bigger on disk, but don't need to be transcoded.

The import options and the corresponding approximate sizes on disk are as follows:

- *ISO*: A literal copy of the entire disk; up to 8.5GB

- *Perfect*: A literal copy of just the track selected; around 3.5GB in my example disk

- *Good*: A transcoded version of just the track selected, at a slightly lower resolution; around 1.2GB for my example disk

For reference, a dual-layer DVD holds around 8.5GB when full.

First off, you need to install the dependencies for mtd. Do this by running the following command:

```
$ sudo apt-get install transcode mplayer
```

This installs transcode, which is a generic video transcoding application capable of reading and writing in many formats. You can find out more about it at http://www.transcoding.org. It also installs mplayer, which is used to get the title and subtitle of the disks. Next, you need to configure the import functionality. You do this under Utilities / Setup ➤ Setup ➤ Media Settings ➤ DVD Settings ➤ Rip Settings. First off, you configure where to store temporary files (which can be quite large), the commands to use to get the title and subtitle of the disk, and the transcode command, all of which you should be able to leave as the defaults (see Figure 12-4).

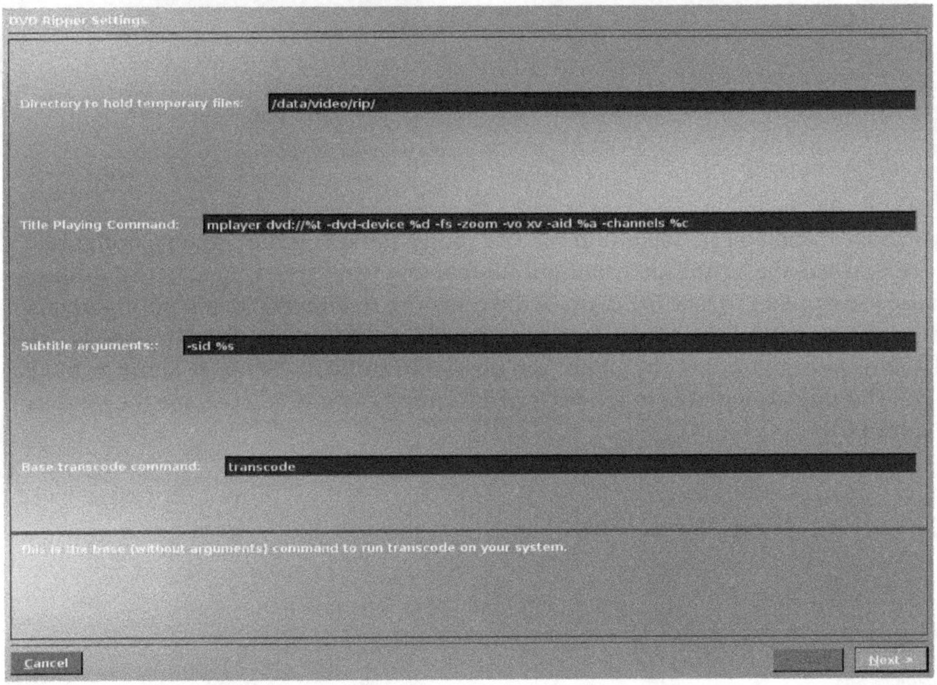

Figure 12-4. *Configuring DVD import*

Don't forget to create that temporary directory. Next, you configure mtd; all the defaults should be correct (see Figure 12-5).

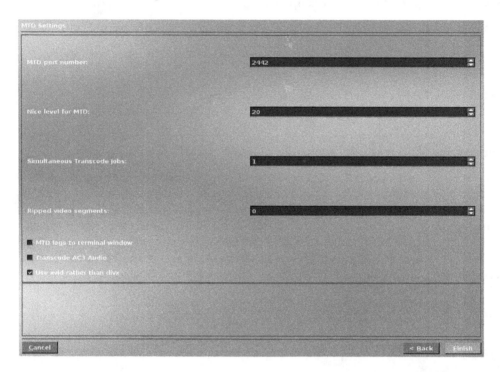

Figure 12-5. *Configuring mtd*

Once you have installed transcode and mplayer, and have set up the ripping configuration, you need to tweak the MythVideo configuration before you're ready to start mtd running. This is because the ripped DVDs will end up in the directory that MythVideo is configured to check for videos to play using the MythVideo plug-in, which means that directory needs to be configured before you can import any DVDs. You do that in Utilities / Setup ➤ Setup ➤ Media Settings ➤ DVD Settings ➤ Video Settings ➤ General Settings, where you will see the textbox shown in Figure 12-6.

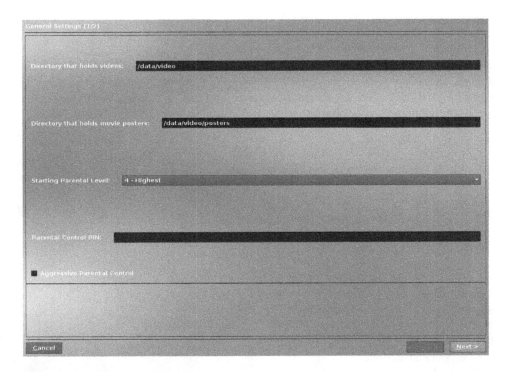

Figure 12-6. *Configuring MythVideo*

You can read more about MythVideo in Chapter 9. Then, just click Next through the rest of the configuration options. At last, you're ready to start mtd running. For the moment, do this interactively (in a terminal window) so that you can see what's happening. Here's an example:

```
$ mtd -n
2006-10-05 15:55:41.195 Using runtime prefix = /usr/local
2006-10-05 15:55:41.429 New DB connection, total: 1
2006-10-05 15:55:41.527 Connected to database 'mythconverg' at host: localhost
mtd started at Thu Oct 5 15:55:41 2006
mtd is running on a host called mikal-laptop
15:55:41: Waiting for connections/jobs
15:55:41: mtd is listening on port 2442
libdvdread: Using libdvdcss version 1.2.5 for DVD access
''15:55:47: DVD inserted: SAMPLE_MOVIE_PAL_AUS
15:55:47:             : Title 1 is of type 6 (dvdinput table)
15:55:47:             : Title 2 is of type 8 (dvdinput table)
15:55:47:             : Title 3 is of type 8 (dvdinput table)
15:55:47:             : Title 4 is of type 8 (dvdinput table)
15:55:47:             : Title 5 is of type 8 (dvdinput table)
15:55:47:             : Title 6 is of type 8 (dvdinput table)
15:55:47:             : Title 7 is of type 6 (dvdinput table)
15:55:47:             : Title 8 is of type 0 (dvdinput table)
15:55:47:             : Title 9 is of type 6 (dvdinput table)
```

```
15:55:47:              : Title 10 is of type 8 (dvdinput table)
15:55:47:              : Title 11 is of type 6 (dvdinput table)
```

You can see here that the transcoding daemon started, and then began scanning the disk—this disk information is what we needed to install mplayer for. You probably just want to leave mtd running as a daemon, though, because it's not interactive. You can set up mtd to start on startup by editing the init script created in Chapter 3, which is stored at /etc/init.d/mythbackend:

```
'! /bin/sh
#
# mythbackend, also starts the MythTV transcode daemon
#

PATH=/usr/local/sbin:/usr/local/bin:/sbin:/bin:/usr/sbin:/usr/bin

BACKEND=/usr/local/bin/mythbackend
BACKEND_NAME=mythbackend

MTD=/usr/local/bin/mtd
MTD_NAME=mtd

MYTHUSER="myth"

test -x $BACKEND || exit 0
test -x $MTD || exit 0

set -e

case "$1" in
  start)
        echo -n "Starting MythTV: $BACKEND_NAME"
        su ${MYTHUSER} -c "$BACKEND -v all -d -l ${MBE_LOGFILE}" &

        echo -n "Starting MythTV: $MTD_NAME"
        su ${MYTHUSER} -c "$MTD -v all -d -l ${MBE_LOGFILE}" &
        echo "."
        ;;
  stop)
        echo -n "Stopping $DESC: $NAME "
        killall $BACKEND
        killall $MTD
        echo "."
        ;;
  *)
        echo "Usage: $0 {start|stop}" >&2
        exit 1
        ;;
esac

exit 0
```

The new parts of the init script are in bold, and the entire script can be downloaded from http://www.mythtvbook.com/, if you prefer.

You're now ready to start archiving the disk. If you left mtd running from the previous example, when you insert a DVD you'll get something like what is shown in the preceding sample mtd output. You then go into the MythTV user interface and navigate to Optical Disks ➤ Import DVD. If you don't have mtd running, then you'll see the screen shown in Figure 12-7.

Figure 12-7. *If you see this message, then you need to start mtd.*

Start mtd as discussed earlier, and then press any number key. Once that is fixed, or if mtd is already running, then you'll see the screen in Figure 12-8 if you haven't got a DVD inserted.

Insert a DVD and press a number key, and you'll end up at the screen shown in Figure 12-9.

Figure 12-8. *If you see this screen, then you need to insert a DVD.*

Figure 12-9. *Selecting the tracks to archive*

Here you can see that track 1 is currently listed as being archived, and is around 1 hour, 36 minutes. Normally, the main content on the disk will be track 1, and it will be the longest track as well. You can navigate through the tracks on the disk using the left and right arrow keys. Use the up and down arrow keys to select the quality item, and then use the spacebar or the OK key on the remote to select the quality. Finally, press the 0 key to start processing. For an ISO or perfect copy, there is no transcode process; otherwise, archival runs in two stages. The first stage (which is the only stage to run for the ISO and perfect copies) is the copying to disk, which looks like Figure 12-10.

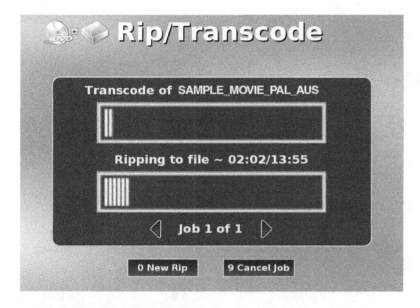

Figure 12-10. *Copying a DVD to disk*

After that, the transcode process will start working out what to do if you're making a transcoded archive (see Figure 12-11).

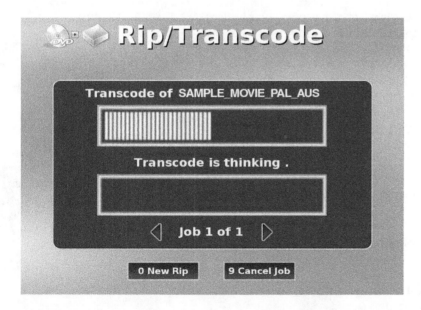

Figure 12-11. *Preparing to transcode*

After a short time, the transcode job will start. These jobs are quite slow, so be patient (see Figure 12-12).

Figure 12-12. *Transcoding*

Once the archive job is complete, there will be an AVI file in your MythVideo location, which you can then play with MythVideo or a variety of other video players. The output from mtd during this transcode process will be something along these lines:

```
04:37:27: launching job: job dvd 1 1 9 0 -1 /var/video/SAMPLE_MOVIE_PAL_AUS
2006-10-06 04:37:28.816 New DB connection, total: 2
2006-10-06 04:37:28.817 Connected to database 'mythconverg' at host: localhost
04:37:28: transcode command will be: transcode -i ➥
/var/video/rip/SAMPLE_MOVIE_PAL_AUS/vob/ -g 720x576 -f 0,3 -M 1 -V -j 0,16,0,16 ➥
-B 5,0 -y xvid -w 1618 -o /var/video/SAMPLE_MOVIE_PAL_AUS.avi --print_status 20 ➥
--color 0
04:37:33: job thread beginning to rip dvd title
libdvdread: Using libdvdcss version 1.2.5 for DVD access

libdvdread: Attempting to retrieve all CSS keys
libdvdread: This can take a _long_ time, please be patient

libdvdread: Get key for /VIDEO_TS/VIDEO_TS.VOB at 0x00000402
libdvdread: Elapsed time 0
libdvdread: Get key for /VIDEO_TS/VTS_01_0.VOB at 0x00000508
libdvdread: Elapsed time 0
libdvdread: Get key for /VIDEO_TS/VTS_01_1.VOB at 0x0000bb3e
libdvdread: Elapsed time 0
libdvdread: Get key for /VIDEO_TS/VTS_02_0.VOB at 0x00342a96
```

```
libdvdread: Elapsed time 0
...snip...
libdvdread: Get key for /VIDEO_TS/VTS_11_0.VOB at 0x00348f8f
libdvdread: Elapsed time 0
libdvdread: Get key for /VIDEO_TS/VTS_11_1.VOB at 0x00348f94
libdvdread: Elapsed time 1
libdvdread: Found 11 VTS's
libdvdread: Elapsed time 2
About to run "transcode -i /var/video/rip/SAMPLE_MOVIE_PAL_AUS/vob/ -g 720x576 ➦
-f 0,3 -M 1 -V -j 0,16,0,16 -B 5,0 -y xvid -w 1618 ➦
-o /var/video/SAMPLE_MOVIE_PAL_AUS.avi --print_status 20 --color 0"
with workdir = /var/video/rip/SAMPLE_MOVIE_PAL_AUS
05:11:25: job thread finished ripping dvd title
2006-10-06 11:06:16.909 Deleting file: ➦
/var/video/rip/SAMPLE_MOVIE_PAL_AUS/vob/SAMPLE_MOVIE_PAL_AUS.vob
11:06:46: job finished successfully: job dvd 1 1 9 0 -1 ➦
/var/video/SAMPLE_MOVIE_PAL_AUS
```

These messages are all informational, but show the process of archiving a DVD. First, the disk is decoded using libdvdread, copied to the hard disk, and then transcoded, before being saved to its final location. This is why the temporary storage location has to be large—it will contain complete copies of the DVD while the transcoding process is happening.

Creating DVDs

Finally in this chapter, we're going to discuss how to create your own DVDs with MythTV. The new, improved way of burning DVDs with MythTV is with MythArchive, which is a new DVD burning implementation intended to replace the older MythBurn. The only real advantage to MythBurn is that is has a web interface (although there is currently work underway to create a web interface for MythArchive, as well).

First of all, you need to install the dependencies (and there are quite a number of them):

```
$ sudo apt-get install ffmpeg dvdauthor dvd+rw-tools udftools mjpegtools mkisofs➦
transcode
```

Next, you need to configure the archival options in Utilities / Setup ➤ Setup ➤ Media Settings ➤ Archive Files Settings. On the first configuration screen, you set up the temporary directory for archiving working files (once again, this needs to be on a file system with a fair bit of free space, as these working files can be quite large), the MythArchive share directory (which is where MythArchive is installed—the default should be fine in this case), the video format for the DVD, some simple rules for which file types can be put onto the DVD, and the location of your DVD burner (which in most cases should be /dev/dvd, but in my case is something different). A sample of this screen is shown in Figure 12-13.

Figure 12-13. *Setting up MythArchive*

NTSC, PAL, AND SECAM

The pictures displayed by your analog television are encoded when transmitted on the cabling used by your various entertainment devices. These encodings are also often used to distinguish the actual media, such as video cassettes and DVDs, although this is not a strictly accurate description of what is occurring (refer to the Wikipedia page on PAL in the following list for more information on this).

Unfortunately, there are three competing encodings, which makes it much harder to use audiovisual equipment and media from other countries. The three competing standards are as follows:

- NTSC (National Television System(s) Committee) is the encoding used in North America, Japan, and Korea. Starting out as a black-and-white format in 1941, it progressed to include color support in 1953. More details of the history of NTSC can be found at `http://en.wikipedia.org/wiki/Ntsc`.

- PAL (Phase Alternation Line) is an encoding used for many other parts of the world. The PAL encoding was introduced in 1967 by Telefunken, and is based on NTSC. PAL offers better video quality than NTSC, which was the original reason for its development. You can read more technical details about PAL on Wikipedia, at `http://en.wikipedia.org/wiki/PAL`.

- SECAM (séquentiel couleur à mémoire) is the encoding used in France, some former communist countries, portions of Africa, and parts of the Middle East. While development was started in 1956, it was not released until 1967; and SECAM televisions were very expensive when released. Many European countries used SECAM at one time, but most have now converted to PAL. Wikipedia's page on SECAM, at `http://en.wikipedia.org/wiki/SECAM`, contains lots more details.

The next screen of the configuration wizard covers encoding options for the DVDs that will be created. The defaults should be fine for most people's purposes. Figure 12-14 shows a sample of the screen.

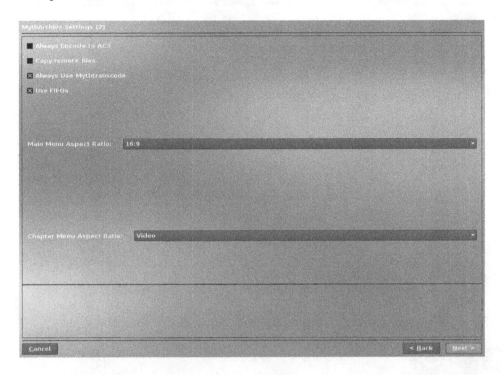

Figure 12-14. *DVD creation options for MythArchive*

Finally, there are two screens that list invocations for various helper commands. You should be fine with the defaults, which are shown in Figures 12-15 and 12-16.

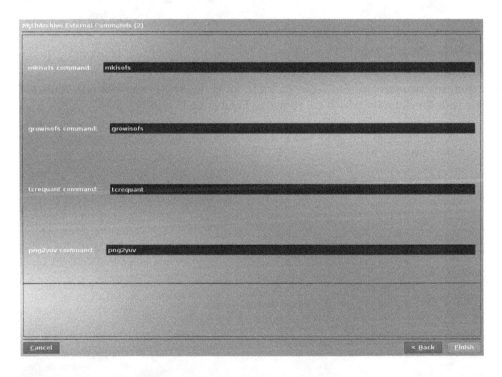

Figure 12-15. *MythArchive command locations, screen 1*

Figure 12-16. *MythArchive command locations, screen 2*

After you have set up MythArchive, you're ready to burn some DVDs. Go to Optical Disks ➤ Archive Files ➤ Find Files to Archive. There you have three options for finding files to write to your DVDs—you can select recordings from television, files from your video collection, or arbitrary files from the file system. Note that the files you select will need to be in a format that MythArchive knows how to convert to the right format for a video DVD, as these are not data disks that you are creating.

For this example, we're going to show you how to burn some videos from MythVideo, although the interface is basically the same for television shows and files from the file system. When you select videos from the menu, you will be presented with a video selection dialog. Figure 12-17 shows my system, which has three videos, all of which are going to be written to the DVD.

Figure 12-17. *Selecting videos to burn to DVD*

The MythArchive interface is a little confusing. The next step is to go backward one menu, to Optical Disks ➤ Archive Files ➤ Export Video Files. You can do two things here: create a DVD or export the files in a native archive. Let's focus on the DVD creation for now, and I'll talk about native archives later in the chapter. Once you've selected Create DVD, you'll be presented with the screen shown in Figure 12-18.

Figure 12-18. *Selecting the media to burn*

The options here, and their corresponding capacities, are as follows:

- *Single layer DVD*: 4.4GB capacity

- *Dual layer DVD*: 8.9GB capacity

- *Rewriteable DVD*: 4.4GB capacity

- *File*: However much space is free on that file system

Your choices here will probably be limited by the DVD burning hardware that you have installed on your machine and the media to which you have chosen to burn. Obviously, the capacity of that media will affect how much video you can fit on the disk, but you can also fit more or less content on the disk by changing the quality of the video on the disk. For example, a low-quality video will take significantly less space on the disk than a high-quality video. (Note that this has to do only with how hard you compress the video—noisy video, which can also be described as lower-quality, can actually take up *more* space when it's compressed.) You'll be asked what quality you would like to use on this disk in a moment.

Now you select the videos that you want to place on this disk. Once the required transcoding has occurred, you'll get feedback about how full the disk is from the capacity graph on the bottom-right-hand side of the screen. In the example shown in Figure 2-19, there are too many videos selected for the disk size.

Figure 12-19. *Selecting the videos to burn to this disk*

The interface here is somewhat unusual, but for a reason. You preselect the videos and files you want to put on disk, and then you select from that subset the videos to put on the current disk. So, if you want to burn ten videos, you can preselect them all, and then only burn the smaller number—which will actually fit onto any given disk. Figure 12-20 shows an example of a selection of videos that will fit on a disk.

Figure 12-20. *Selecting the videos to burn to the disk*

Next, you can select a theme for the DVD menus. This is the user interface that is presented to you when you insert the DVD into your DVD player. The selection interface shows you examples of the options available. Figure 12-21 shows an example.

Figure 12-21. *Selecting a theme for the DVD menus*

There are a variety of themes available, each of which is described briefly in the following list:

- *Simple*: This theme lacks extraneous graphics and autoplays the videos in sequence with no menu and no introduction (see Figure 12-22).

Figure 12-22. *The simple theme*

- *Compact*: This theme has an introduction video and contains six videos per menu page. There is no scene selection menu, and the disk does not autoplay videos (see Figure 12-23).

Figure 12-23. *The compact theme*

- *MythCenter/MythCenter autoplay*: This theme has an introduction video and displays three videos per main menu page. Each video is broken into eight chapters, with a chapter selection menu. The autoplay version also autoplays the videos in sequence (see Figure 12-24).

Figure 12-24. *The MythCenter theme*

- *G.A.N.T.*: This theme has an introduction video and three videos per main menu page. Each video is broken into eight chapters, with a chapter selection menu. There is a program details page for each video as well (see Figure 12-25). The difference between this theme and the MythCenter theme is the background graphics and the look and feel of the user interface.

Figure 12-25. *The G.A.N.T. theme*

It's also possible to create your own MythArchive themes. Check out /usr/local/share/mythtv/ mytharchive/themes/G.A.N.T./ for an example of a well-implemented theme. Once you've selected a theme, you need to select the encoder profile to use for the DVD (see Figure 12-16). (This is where the varying quality of storage mentioned earlier in chapter comes into play.)

Figure 12-26. *Selecting an encoder profile*

Your choices here are as follows:

- *SP (standard playback quality)*: This option gives you two hours of video on a single-layer DVD.

- *LP (long play)*: This option gives you four hours of video on a single-layer DVD, but at a lower quality than a standard playback disk.

- *EP (extended play)*: This option gives you six hours of video on a single-layer DVD, but with a further decrease in playback quality.

- *HQ (high quality)*: This option gives you only an hour of video on a disk, but at a higher than normal quality.

- *Don't re-encode*: This options lets you just write whatever video format is stored by MythTV to the DVD. This might result in disks that cannot be played on DVD players.

Select one of these options. Once again, the meter on the bottom-right will indicate if you have too many videos for the encoding and media combination that you have selected. Finally, encoding and burning will start. A transcoding process will occur before the disk is burned, which can take quite some time. Figure 12-27 shows the dialog that displays the progress.

Figure 12-27. *Encoding and burning progress*

You can click the Exit button to stop watching the progress of the job. The job will continue without you. If you choose to burn another DVD, it will return you to the progress screen. If you want to cancel the job, simply press the Cancel button. If you have a crashed job that stops you from burning another DVD, then manually remove the directories in the MythArchive temporary directory that you created earlier.

Native Archives

We promised to explain native archives, which is the other export option from MythArchive. When you select Create Native Archive from Optical Disks ➤ Archive Files ➤ Export Video Files, you select the media you are archiving to in the same way that you did for the DVD creation process earlier. Then you select videos just like you do when you include them on a DVD. When the disk is produced, it includes all the information needed to restore those recordings to the MythTV database, including metadata such as air time. You import the archives using Optical Disks ➤ Archive Files ➤ Import Video Files, and then navigate through the file system to the archived files.

An archive system like this is great for backing up videos that you don't want to lose, as well as moving recordings between MythTV systems. You should note that nuvexport (discussed in Chapter 5) also has similar functionality. The advantage of MythArchive is that it has a graphical user interface that integrates with the rest of MythTV, while nuvexport is easy to script for automatic backup and export jobs.

Conclusion

In this chapter, we've discussed the DVD functionality in MythTV, from playback of DVDs to creating your own disks. While DVD playback is pretty much essential functionality, the DVD creation functionality now available with MythArchive is above and beyond what most PVRs have, and is an good example of why the flexibility of MythTV makes it extraordinary.

In the next chapters, we're going to cover some final topics around MythTV—how to write a simple on-screen display script, how to use the Internet telephony plug-in, and how to run the absolute latest versions of MythTV—so keep reading.

■ ■ ■

## CHAPTER 13

Controlling MythTV over the Network and On-Screen Displays

Despite the large list of MythTV features discussed already in this book, we want to talk about two more. Those features are on-screen display, which gives you the ability to display arbitrary text on the screen of your MythTV frontends during video playback, and Network Control, which lets you control a frontend over the network. Unfortunately, the on-screen display doesn't work if the MythTV machine is displaying a menu or playing a video or DVD (unless you use the internal player). However, it is still a useful feature.

Most of the chapters in this book have focused on how to get a specific piece of MythTV working nicely. This chapter is a little different. Although we will show you how to just use the on-screen display utility, `mythtvosd`, we'll also share some project ideas for cool things to do with it; in addition, we'll walk you through an example implementation of one of those ideas. This project also uses the Network Control interface, which is why it is in this chapter as well.

The project is a Google Talk/Jabber interface to MythTV, which lets you send commands via instant messaging to your MythTV frontend. It implements all the functionality available in the MythTV `telnet` Network Control (Telnet) interface except for the online help, as well as on-screen displays (which are not implemented in the Network Control interface). The implementation will involve reading some code, but don't feel daunted by that. If you're not a programmer, you'll still be able to download the source code for the finished product and give it a go.

We've also arranged the code so that the portion that actually uses MythTV is as small as possible and separate from the infrastructure needed to make it work, which means you can ignore the infrastructure code if you want.

First you'll learn how to make MythTV display on-screen text from the command line.

Displaying On-Screen Text

MythTV can display on-screen text as long as there is a recording or Live TV playing. In other words, it won't work if you're watching a show with something other than the MythTV internal player, and it won't work if you're currently in the menu system. With those limitations in mind, it's actually a useful feature. Common uses for on-screen displays include displaying caller ID information for incoming phone calls, warning the user about error conditions such as being low on disk space, and informing the user of waiting email. Figure 13-1 shows an example of an on-screen display.

Figure 13-1. *A sample of an on-screen display*

The on-screen display utility, mythtvosd, uses templates to display the text and define how the on-screen display will look. Templates are specified with the --template command-line option. The previous example was made with this command:

```
mythtvosd --template=alert alert_text='This is some sample text'
```

We'll show you some of the templates and their display options in a little while, but the important thing to note is that the name of the alert_text argument changes based on the template, which means that if you write scripts using mythtvosd, then you're going to need to take that into account.

mythtvosd works by sending a UDP packet over the network, so you can also send the request to display an on-screen message to a remote frontend or to all the frontends on a given network broadcast region. Change the IP address used from the default of localhost (127.0.0.1) using the --bcastaddr command-line option. You can also change the port that the UDP packet is sent to with the --udpport command-line option, but that's less useful.

Using Templates

mythtvosd implements three main templates for on-screen displays. The following sections show you examples of these three templates, with sample output.

alert

The alert template has already been shown in this chapter. Figure 13-2 shows another sample.

Figure 13-2. *The* alert *template*

Table 13-1 lists the only substitution.

Table 13-1. alert *Template Substitutions*

Substitution Text	Use For
alert_text	The text to display

Here's a sample:

```
mythtvosd --template=alert alert_text='This is some sample text'
```

cid

The cid template is aimed at displaying caller ID information from computerized phone systems (see Figure 13-3).

Figure 13-3. *The* cid *template*

It has the substitutions listed in Table 13-2.

Table 13-2. cid *Template Substitutions*

Substitution Text	Use For
caller_line	If you have more than one phone line that the call could have occurred on, then set this to be the caller line.
caller_name	The name of the caller, if known.
caller_number	The number the caller called from, if known.
caller_date	The date of the call.
caller_time	The time of the call.

Here's a sample:

```
mythtvosd --template=cid caller_line=caller_line caller_name=caller_name ➥
caller_number=caller_number caller_date=caller_date caller_time=caller_time
```

scroller

The scroller template scrolls text from the bottom right to the bottom left of the screen and simulates those news tickers that some television stations use (see Figure 13-4).

Figure 13-4. *The* scroller *template*

It has the substitutions listed in Table 13-3.

Table 13-3. scroller *Template Substitutions*

Substitution Text	Use For
scroll_text	The text to scroll across the bottom of the screen

Here's a sample:

```
mythtvosd --template=scroller scroll_text='This is some sample text'
```

Controlling a MythTV Frontend over a Network

The 0.19 release of MythTV introduced another frontend interface, called Network Control, to control a running MythTV frontend over a network. Why would you want to do this?

- You could write a little program to manipulate the frontend, such as when an event happens. For example, if the phone is ringing, the TV pauses.

- Use your laptop as a (albeit expensive) remote!

- Monitor what somebody is watching on the frontend (to stop the TV and force people to come to dinner).

Currently, in 0.20, Network Control is quite useful. You can basically do anything you can do with the remote control.

Enabling Network Control

In mythfrontend, the General Setup menu contains an option to enable Network Control. This will make the frontend listen on port 6546. You need to enable this option only once, because mythfrontend will remember the option. Now, when you start mythfrontend, the log messages will also mention that Network Control is enabled and listening on port 6546.

Using Network Control

You connect to the MythTV frontend using a telnet client. Telnet has no security, and the MythTV frontend provides no password authentication (even if it did, it would be easy to break since Telnet has no encryption of the data it sends over the network). Since your MythTV box won't be connected directly to the Internet (usually you'll have it behind a router of some sort), this isn't a problem.

```
$ telnet localhost 6546
Trying 127.0.0.1...
Connected to localhost.localdomain.
Escape character is '^]'.
MythFrontend Network Control
Type 'help' for usage information
---------------------------------
# help
Valid Commands:
---------------
jump            - Jump to a specified location in Myth
key             - Send a keypress to the program
play            - Playback related commands
query           - Queries
exit            - Exit Network Control

Type 'help COMMANDNAME' for help on any specific command.

# help query
query location      - Query current screen or location
query recordings    - List currently available recordings
query recording CHANID STARTTIME
                    - List info about the specified program
```

```
# query location
Playback Recorded 23:11 of 29:58 1x 1010 2006-05-21T14:00:00
```

Using MythBot, the IM Interface to MythTV

Now we'll show you some sample code that uses both the MythTV frontend Network Control
interface and `mythtvosd` to implement some cool functionality. We'll start by showing you the
specific code that implements the MythTV functionality, which in this case is a `gtalkbot` module called MythBot. If you don't have some experience in Python programming, then this section
will look somewhat mysterious to you. Rest assured, it's still cool.

     `gtalkbot` works by allowing modules to register verbs that they will implement. A *verb* is
the first word of an instant message and is used to define what command is going to be executed. MythBot registers all the commands that are available in the frontend Network Control
interface, with the exception of the `help` command, as verbs. It also registers the word say as
a verb for use with the MythTV on-screen display functionality.

     Then, when `gtalkbot` sees one of these verbs, it calls MythBot and asks for the command
to be executed. This is relatively simple because all the commands are just handed to the
frontend Network Control interface, unless it is the say verb, which executes `mythtvosd`.

■**Note** The code we'll show you here is a slightly old version of `gtalkbot`. We're showing you an old version because it's easier to understand this version, and it demonstrates the concepts without getting bogged
down in complex implementation details (such as configuration files). You should check for a new version of
`gtalkbot` at `http://www.stillhq.com/gtalkbot/` before actually installing it.

     The following is the code for the module, which is written in Python. This first block is
simple. It defines that this is a Python 2.4 script, contains a simple documentation comment
describing what the module does, and then imports the Python libraries that the module needs
to operate:

```
#!/usr/bin/python2.4
#
# Copyright (c) Michael Still 2006, released under the terms of the GNU GPL v2

"""mythtv.py -- a MythTV gtalkbot module

This module implements the MythTV on-screen display and telnet functionality
in a simple example gtalkbot bot.
"""

import sys
import socket
import os

mythtv = None
```

The next block of code starts the MythTelnet class, which implements the network connection to the MythTV frontend Network Control interface. It does this by opening a socket connection to the right port on the local machine in its __init__ method.

```
class MythTelnet:
  """Execute commands using the MythTV frontend telnet interface"""

  def __init__(self):
    try:
      self.s = socket.socket(socket.AF_INET, socket.SOCK_STREAM)
      self.s.connect(('localhost', 6546))
      self.s.recv(1024)
    except:
      print 'Connection failed for MythTV telnet interface'
      sys.exit(1)
```

The following is the other method in the MythTelnet class. It provides a simple method for users of the class to execute a command via the Network Control interface by simply passing the command as a string to the method. It executes the command by writing the command to the socket and then reading until it is sure that it has the entire response to the command, which is then returned as a single string. It is possible that the reply to a given command might arrive in more than one read, which is why this code needs to loop until there is no more to read.

```
def execute_command(self, cmd):
    """Execute text commands via the frontend telnet interface, and return
    the string output from the command"""

    self.s.send(cmd + '\n')
    result = ''
    data = ''

    while data[-2:] != '# ':
      result = result + data
      data = self.s.recv(1024)
    result = result + data[:-3]
    return result.rstrip('\r\n')
```

The following method is not in a class. The Verbs() method is called by gtalkbot when the module is loaded. It is the module's chance to register the verbs that it would like to own in the instant message interface.

```
def Verbs():
    """Verbs -- return the verbs which this module supports

    Takes no arguments, and returns an array of strings.
    """

    return ['jump', 'key', 'play', 'query', 'say']
```

In this case, the following verbs are implemented:

- jump

- key

- play

- query

- say

These verbs should be all in lowercase, because gtalkbot forces verbs into lowercase before checking for a match on input. All these verbs are implemented by calling the execute_command method in the MythTelnet class, apart from the say verb.

The following method is called by gtalkbot when it needs to decide what to display as the status message for the bot user. In reality, it will generally ask for a suggested status message after a verb from that module has been used. The method returns a string, which in the case of MythBot is always the current location in the user interface.

```
def Status():
    """Status -- suggest a message to display as the status string for the bot

    Takes no arguments, and returns a string. In this case we return the current
    location in the MythTV user interface.
    """

    return mythtv.execute_command('query location')
```

The Command method is called when gtalkbot detects that the user has sent a message that starts with a verb that matches one registered by the MythBot module. The arguments are the verb that was detected and the remainder of the instant message. If the verb is say, then the remainder of the message is displayed on the frontend using the mythtvosd command. Otherwise, the command is sent to the MythTV frontend Network Control interface.

```
def Command(verb, line):
    """Command -- execute a given verb with these arguments

    Takes the verb which the user entered, and the remainder of the line.
    Returns a string which is sent to the user.
    """

    if verb == 'say':
        if mythtv.execute_command('query location')[:17] == 'Playback Recorded':
            os.system('mythtvosd --template=alert alert_text="' + line + '"')
            return 'Done'
        else:
            return 'Unable to say, because frontend is not playing'

    # If it's not a say, then send it on to the MythTV frontend telnet interface
    return mythtv.execute_command(verb + ' ' + line)
```

This method is called when the module is about to be unloaded, and it lets you do any cleanup that might be required. In this case, no cleanup is needed:

```
def Cleanup():
  """Cleanup -- you're about to be unloaded.

  We don't need to do anything in this case.
  """

  return
```

Finally, the following code is executed when the module is loaded. It simply creates an instance of the MythTelnet class:

```
# Do initialization here
print u'mythtv_bot is loading'
mythtv = MythTelnet()
```

Using gtalkbot, the Modular IM Engine

The gtalkbot code is reasonably simple as well when you take into account that it has to connect to the Google Talk instant messaging network. If you're not interested in the bot engine, then feel free to skip this section and move on to the next. The code is a little long as well, so we'll try to cover the important points without getting too bogged down in the details.

The first block of code is the documentation comment for the Python script and the import of the various Python libraries needed to implement gtalkbot:

```
#!/usr/bin/python2.4
#
# Copyright (c) Michael Still 2006, released under the terms of the GNU GPL v2

"""gtalkbot.py -- a modular Jabber bot

This is a simple Python Google Talk and Jabber bot. It loads modules from
the specified module directory, and then lets them register "verbs." These
verbs are simple single-word commands that must be at the start of the line
entered by the other side of the chat conversation. If the first word of one
of these lines matches a registered verb, then the relevant module is
handed the verb, as well as the remainder of the line (in separate arguments).

The bot authenticates clients before it lets any verbs be executed.

This code is very heavily based on the echobot example from the PyXMPP
distribution.
"""


import sys
import logging
import locale
```

```
import codecs
import os
import re
import imp

from pyxmpp.all import JID, Iq, Presence, Message, StreamError
from pyxmpp.jabber.client import JabberClient
from pyxmpp import streamtls

modules = {}
verbs = {}
```

The PasswordCache class loads a list of passwords from /usr/local/share/gtalkbot/passwd. The format of the file is simple; each line contains the Google Talk username of a user, a tab, and then a plain-text password. The password is stored in plain text on disk, because it is also sent to the bot via a plain-text instant message (although messages are encrypted on the network if you are using Google Talk or a Jabber client and server with SSL support). Finally for this class, the authenticate method is used to verify that a given username and password pair appears in the password file.

Note This behavior has changed in newer versions of gtalkbot. See http://www.stillhq.com/gtalkbot/ for more details.

```
class PasswordCache:
  """Reads a file of jabber IDs and passwords at startup, and then provides
    lookups for those for authentication"""

  def __init__(self):
    """Read /usr/local/share/gtalkbot/passwd into a dictionary."""
    self.passwd = {}

    # We load passwords from a hard coded path for this example, but in a
    # real application you would do this better...
    print u'Loading passwords'
    try:
      passwdfile = open('/usr/local/share/gtalkbot/passwd', 'r')
    except:
      print u'Please create the password file at /usr/local/share/gtalkbot/passwd'
      sys.exit(1)

    for line in passwdfile.readlines():
      [jid, password] = line.rstrip('\r\n').split('\t')
      self.passwd[jid] = password
```

```
def authenticate(self, jid, password):
  """Determine if a given user has provided a password which matches the
  cache"""

  if not self.passwd.has_key(jid):
    return False;
  print u'Authenticating %s with "%s" against "%s"' %(jid, password,
                                          self.passwd[jid])
  return self.passwd[jid] == password
```

The Client class implements all the Google Talk or Jabber parts of gtalkbot. This is done by inheriting from the PYXMPP JabberClient class, which implements an event loop for events that can occur in a Google Talk or Jabber session. The dictionary called authenticated here is used to store a list of the usernames that have previously been authenticated and is initially empty.

```
class Client(JabberClient):
  """This class implements all the Jabber parts of the program, a lot
  of the Jabber functionality comes from inheriting from JabberClient,
  which is part of PyXMPP."""

  authenticated = {}
```

The following code connects to the correct Google Talk or Jabber server for the user who the bot is running as:

```
def __init__(self, jid, password):
    # If a bare JID is provided add a resource
    if not jid.resource:
      jid=JID(jid.node, jid.domain, "gtalkbot")

    # Setup client with provided connection information and identity data,
    # this block also handles the SSL encryption using with Google Talk
    tls = streamtls.TLSSettings(require=True, verify_peer=False)
    auth = ['sasl:PLAIN']
    JabberClient.__init__(self, jid, password,
                          disco_name="gtalkbot", disco_type="bot",
                          tls_settings=tls, auth_methods=auth)

    # Register features to be announced via Service Discovery
    self.disco_info.add_feature("jabber:iq:version")
```

Finally, you initialize the password cache for the bot. Note the side effect here that the password file is loaded only on start-up. If a password is changed or a user is added to the password file, then the bot needs to be restarted before that change will take effect. This isn't a big problem because the bot is stateless apart from the initial authentication of users, so other users will simply notice that the bot goes offline briefly during the restart and that they then need to reauthenticate.

```
# Initialize the client
print u'Initialize client'
self.passwd = PasswordCache()
```

The following method registers all the methods that should be called when various events occur in the client's event loop. Essentially this is all about routing events to the right methods and isn't very interesting.

```
def session_started(self):
    """This is called when the IM session is successfully started
    (after all the necessary negotiations, authentication, and
    authorization)."""
    JabberClient.session_started(self)

    # Set up handlers for supported <iq/> queries
    self.stream.set_iq_get_handler('query', 'jabber:iq:version',
                                   self.get_version)

    # Set up handlers for <presence/> stanzas
    self.stream.set_presence_handler('subscribe', self.presence_control)
    self.stream.set_presence_handler('subscribed', self.presence_control)
    self.stream.set_presence_handler('unsubscribe', self.presence_control)
    self.stream.set_presence_handler('unsubscribed', self.presence_control)

    # Set up handler for <message stanza>
    self.stream.set_message_handler('normal', self.message)
```

When the Google Talk or Jabber server requests the version number of the client that is connected, then the event will get routed to this function, which returns that the client is gtalkbot version 1.0:

```
def get_version(self,iq):
    """Handler for jabber:iq:version queries.

    jabber:iq:version queries are not supported directly by PyXMPP, so the
    XML node is accessed directly through the libxml2 API.  This should be
    used very carefully!"""

    iq = iq.make_result_response()
    q = iq.new_query('jabber:iq:version')
    q.newTextChild(q.ns(), 'name', 'gtalkbot')
    q.newTextChild(q.ns(), 'version', '1.0')
    self.stream.send(iq)
    return True
```

The following is the method that is called every time a instant message is received by gtalkbot. This first block of code extracts the body, sender, subject, and type of the message received so that you can decide what to do with the message:

```
def message(self, stanza):
    """Handle incoming messages"""

    subject = stanza.get_subject()
    body = stanza.get_body()
    sender = stanza.get_from().as_utf8().split('/')[0]
    t = stanza.get_type()
```

If the message is a text message (as opposed to a new status message or a presence message, for example), then determine whether the user is authenticated. If the user isn't, then this message must contain an auth verb. If the user is authenticating (that is, this line starts with the auth verb), then determine whether the password provided matches the one in the cache. If it doesn't, then let the user know to try again. If the verb is anything other than auth, then let the user know that they need to authenticate before they can do anything using the bot.

```
    # If this is a text message, then we respond
    if stanza.get_type() != 'headline' and body:
        verb = body.split(' ')[0]
        verb = verb.lower()

        status_message = ''

        # Force users to be authenticated before they can do anything
        if not self.authenticated.has_key(sender):
            if verb == 'auth':
                if self.passwd.authenticate(sender, body[5:]):
                    result = 'Welcome'
                    status_message = 'Authenticated'
                    self.authenticated[sender] = True
                else:
                    status_message = 'Who are you?'
                    result = 'Sorry, try again'
            else:
                status_message = 'Who are you?'
                result = 'You need to authenticate.\nUse auth <password>'
```

As the comment indicates, if the user has authenticated, then find out which module has that verb registered, and let that module know about the message so it can execute the verb. Once the command is executed, also ask that module whether it has a suggestion for what the bot user's new status message should be.

```
        # If we are authenticated, then other commands work
        elif verbs.has_key(verb):
            module_name = verbs[verb]
            result = modules[module_name].Command(verb, body[len(verb) + 1:])
            status_message = module.Status()
```

If no one has registered this verb, then just let the user know that you don't know what to do with it:

```
else:
    result = 'Command not known'

    status_message = 'Huh?'
```

Each of the following blocks of handling code set the return message for this instant message and the status string that gtalkbot displays to all users. This code now sends the status message and then sends the response to the instant message that was received.

```
p = Presence(status = status_message)
self.stream.send(p)

m = Message(to_jid = stanza.get_from(),
            from_jid = stanza.get_to(),
            stanza_type = stanza.get_type(),
            subject = subject,
            body = result)
self.stream.send(m)
```

Returning True here indicates that you have handled the message:

```
return True
```

When a user asks to be in your buddy list or when a user in your buddy list connects, then just approve the request. We let anyone connect because we ask them to authenticate later.

```
def presence_control(self,stanza):
    """Handle subscription control <presence/> stanzas -- acknowledge
    them."""

    self.stream.send(stanza.make_accept_response())
    return True
```

The following is the setup code for gtalkbot. You need to use Unicode (a text format) for the text output that the bot does, because Google Talk and Jabber messages that arrive are in Unicode. You also need to set up the logging engine that the PYXMPP library uses for its logging.

```
# The XMPP protocol is Unicode-based. To properly display data received
# it _must_ be converted to local encoding or UnicodeException may be
# raised
locale.setlocale(locale.LC_CTYPE,"")
encoding = locale.getlocale()[1]
if not encoding:
    encoding = 'us-ascii'
sys.stdout = codecs.getwriter(encoding)(sys.stdout, errors='replace')
sys.stderr = codecs.getwriter(encoding)(sys.stderr, errors='replace')
```

```
# PyXMPP uses `logging` module for its debug output applications should
# set it up as needed
logger = logging.getLogger()
logger.addHandler(logging.StreamHandler())
logger.setLevel(logging.INFO)
```

Finally, for the gtalkbot code, you check that the user provided the correct arguments when starting the engine, load the modules from the right directory, and connect to the instant messaging network:

```
if len(sys.argv) < 4:
  print u'Usage:'
  print '\t%s moduledir JID password' % (sys.argv[0],)
  print 'example:'
  print '\t%s /usr/local/gtalkbot/modules username@gmail.com verysecret' \
    % (sys.argv[0],)
  sys.exit(1)

# Load modules
print u'Loading modules...'
re_module = re.compile('[^.].*\.py$')
for module_file in os.listdir(sys.argv[1]):
  if re_module.match(module_file):
    name = module_file[:-3]
    print 'Loading %s' % name
    module_info = imp.find_module(name, [sys.argv[1]])
    module = imp.load_module(name, *module_info)
    modules[name] = module

# Load the verbs supported by this module
for verb in module.Verbs():
  verbs[verb] = name

print u'Creating client...'
c = Client(JID(sys.argv[2]), sys.argv[3])

print u'Connecting...'
c.connect()

print u'Connected'
try:
  # The client class provides a basic "main loop" for the application.
  c.loop(1)

except KeyboardInterrupt:
  print u'Disconnecting'
  c.disconnect()
```

Running gtalkbot with the MythBot Module

The last thing we'll show you in this chapter is how to start the `gtalkbot` engine (including all the things you need to do to set it up) and then show a simple example session. We'll also provide a summary of the list of commands that are supported at the moment. Let's start off by setting up the engine and the MythBot module.

Note There are possibly newer versions of `gtalkbot` than what we discuss in this chapter. Check `http://www.stillhq.com/gtalkbot/` for newer versions before following these instructions.

Installing the Code

You can download the `gtalkbot` code, which includes the MythBot module, from `http://www.stillhq.com/gtalkbot/`. Once you've downloaded it, extract it from the tarball to wherever you want to install it. We chose to install our copy in `/usr/local/share/`, like this:

```
$ sudo mkdir -p /usr/local/share/gtalkbot/
[...You'll need to change permissions on the directory using chown ➡
so that you can access the directory...]
$ cd /usr/local/share/gtalkbot/
$ wget http://www.stillhq.com/gtalkbot/source/gtalkbot-latest.tgz
$ tar xvzf gtalkbot-latest.tgz
```

That's all the installation you need.

Installing the Dependencies

You also need to install some packages to get PYXMPP working:

```
$ sudo apt-get install python2.4-pyxmpp
```

Configuring gtalkbot

The only configuration needed is to provide a password file for `gtalkbot`. `gtalkbot` expects that file to be at `/usr/local/share/gtalkbot/passwd`. The format is a simple one line per user, with the username of the user who is connecting, a tab, and then the password for that user in plain text. The password is stored in plain text on disk, because it is also sent to the bot via a plain-text instant message (although messages are encrypted on the network if you are using Google Talk or a Jabber client and server with SSL support). Here is an example of a password file:

```
mikalguy@gmail.com          gerkin
mikal@otherdomain.com    banana
```

Giving It a Try

The MythBot module assumes that it will be running on the same machine as the MythTV front-end that you want to control, so make sure you run gtalkbot on the right machine. You'll need a Google Talk or Jabber username for the bot itself, but these are relatively easy to get. To sign up for a Google Talk account for the bot, just create a Gmail account at https://www.google.com/accounts/NewAccount?service=talk. Once you know the username and password to use, run gtalkbot like this:

```
$ ./gtalkbot.py modules tvuser@gmail.com password
Loading modules...
Loading mythtv
mythtv_bot is loading
Creating client...
Initialize client
Loading passwords
Connecting...
Connected
```

gtalkbot won't exit, so just leave it running and handling messages. If you get the number of arguments wrong, then you will be presented with a usage message like this:

```
$ ./gtalkbot.py
Usage:
        ./gtalkbot.py moduledir JID password
example:
        ./gtalkbot.py /usr/local/gtalkbot/modules username@gmail.com verysecret
```

If the frontend isn't running on the local machine or has the Network Control functionality turned off, then you'll get an error message like this:

```
$ ./gtalkbot.py modules username@gmail.com verysecret
Loading modules...
Loading mythtv
mythtv_bot is loading
Connection failed for MythTV telnet interface
```

You can find out how to enable the Telnet Network Control interface for MythTV's frontend in Chapter 4. We'll now show you a session with gtalkbot and the MythTV module so you can get a feel for what is possible (see Figure 13-5). For these figures, we've used an open source instant message application called Gaim.

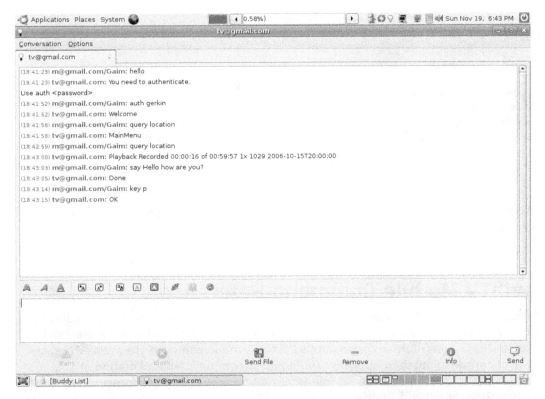

Figure 13-5. *A sample* gtalkbot *session using the MythTV module*

Here is the same session as text, with explanations of what is happening included. What we typed is in bold:

```
(18:41:23) m@gmail.com/Gaim: hello
(18:41:23) tv@gmail.com: You need to authenticate.
Use auth <password>
```

As mentioned earlier, gtalkbot requires users to enter a password before they can use the bot. If they are not authenticated and the verb they send is not auth, then they get an error message until they authenticate.

```
(18:41:52) m@gmail.com/Gaim: auth gerkin
(18:41:52) tv@gmail.com: Welcome
(18:41:58) m@gmail.com/Gaim: query location
(18:41:58) tv@gmail.com: MainMenu
```

Once authenticated, you ask the MythTV module where in the user interface the frontend is currently. In this case, it's displaying the main menu (the top level menu of the frontend). You then used the remote control to start a recording playing:

```
(18:42:59) m@gmail.com/Gaim: query location
(18:43:00) tv@gmail.com: Playback Recorded 00:00:16 of 00:59:57 1x 1029 ➥
2006-10-15T20:00:00
```

```
(18:43:03) m@gmail.com/Gaim: say Hello how are you?
(18:43:05) tv@gmail.com: Done
(18:43:14) m@gmail.com/Gaim: key p
(18:43:15) tv@gmail.com: OK
```

Because there is now a recording playing, the result of the location query now indicates that a video recorded on October 15, 2006, is playing and that we're 15 seconds into the 59 minutes of the video. Incidentally, we're playing the video at its regular speed. Now that we're playing a video, the say verb will work, so we use it to display some text via the MythTV on-screen display command.

Note Newer versions of MythBot than the one shown in this chapter actually tell you the name of the recording being played, not just when it was recorded.

MythTV Module Command Reference

To make the code for the example in this chapter shorter, we haven't included the online help for the modules. We're therefore including a short command reference for the commands here. Note that all but one of these commands are actually implemented by the MythTV Network Control interface, and therefore you can find more complete documentation at http://www.mythtv.org. The summary in the following sections is just a nicely reformatted version of the help from the network interface.

The jump Verb

This jumps to a specified location. Possible locations include those listed in Table 13-4.

Table 13-4. *Jump Locations*

Jump Location	Use for Jumping To
channelpriorities	Channel recording priorities
channelrecpriority	Channel recording priorities
deletebox	TV recording deletion
deleterecordings	TV recording deletion
flixbrowse	Netflix browser
flixhistory	Netflix history
flixqueue	Netflix queue
guidegrid	Program guide
livetv	Live TV
livetvinguide	Live TV in guide
mainmenu	Main menu

Jump Location	Use for Jumping To
managerecordings	Manage recordings/fix conflicts
manualbox	Manual record scheduling
manualrecording	Manual record scheduling
musicplaylists	Select music playlists
mythgallery	MythGallery
mythgame	MythGame
mythmovietime	MythMovieTime
mythnews	MythNews
mythvideo	MythVideo
mythweather	MythWeather
playbackbox	TV recording playback
playbackrecordings	TV recording playback
playdvd	Play DVD
playmusic	Play music
previousbox	Previously recorded
progfinder	Program finder
programfinder	Program finder
programguide	Program guide
programrecpriority	Program recording priorities
recordingpriorities	Program recording priorities
ripcd	Rip CD
ripdvd	Rip DVD
statusbox	Status screen
videobrowser	Video browser
videogallery	Video gallery
videolistings	Video listings
videomanager	Video manager
viewscheduled	Manage recordings/fix conflicts

The key Verb

This verb is used to send keypress events to the frontend. You can send letters, numbers of keycodes. Keycodes include the backslash, backspace, left bracket, right bracket, colon, down arrow, Enter, equal, Escape, F1, F10, F11, F12, F2, F3, F4, F5, F6, F7, F8, F9, greater-than symbol, left arrow, less-than symbol, Page Down, Page Up, right arrow, semicolon, slash, spacebar, and up arrow.

The play Verb

Use this verb to execute playback-related commands. In Table 13-5, square brackets and pipes are used to show alternatives. For example:

```
play channel [up | down]
```

This means that there are two commands: play channel up and play channel down.

Table 13-5. *Play Commands*

Command	Use For
channel [up | down]	Changing channel up or down
channel NUMBER	Changing to a specific channel number
chanid NUMBER	Changing to a specific channel ID (chanid)
program CHANID yyyy-mm-ddThh:mm:ss	Playing program with chanid and starttime
program CHANID yyyy-mm-ddThh:mm:ss resume	Resuming program with chanid and starttime
seek [beginning | forward | backward]	Seeking to the beginning of the recording, forward or backward
seek HH:MM:SS	Seeking to a specific position
speed [pause | normal]	Pausing playback or returning to normal speed
speed [1x | 2x | 4x | 8x | 16x]	Playing back at normal speed or a multiple of normal speed
speed [1x | 1/2x | 1/4x | 1/8x | 1/16x]	Playing back at normal speed or a fraction of normal speed
stop	Stopping playback

The query Verb

Queries are simple searches. Table 13-6 shows the query options.

Table 13-6. *Query Options*

Query	Use For
location	Querying current screen or location
recordings	Listing currently available recordings
recording CHANID STARTTIME	Listing information about the specified program

The say Verb

You use the say verb to display text on the screen, but it works only when mythtvosd works. In other words, you need to be playing a recording for say to work. However, if the frontend is not playing a recording, you will be warned that your message was not displayed. Use the verb like this:

```
say This is the text to display
```

Don't use quotes and so forth around the text, because this will confuse mythtvosd when it is executed.

Conclusion

In this chapter, you learned how to use the `mythtvosd` command to display text on top of a playback happening on a MythTV frontend. We showed you the templates you can use for different kinds of displays, and we introduced you to `gtalkbot` and its MythTV module, which implements a bunch of instant messaging functionality for MythTV and was written especially for this chapter.

You're getting close to the end of the book now; in the remaining chapters, we'll show you how to implement VoIP (Voice over IP, also known as Internet telephony) using MythTV and how to run the absolute latest version of MythTV from the developers' source control system, as well as how to become an active member of the MythTV community, so read on.

CHAPTER 14

■ ■ ■

MythPhone: Using VoIP with MythTV

MythPhone is the MythTV plug-in that lets you make phone calls over the Internet with MythTV. It also supports making video calls with an attached webcam. It talks the standard SIP protocol, so calling other SIP clients or devices over the Internet won't cost you a cent. Also, many companies offer accounts on their systems that let you call regular phone lines (usually at a much lower rate than big telecommunications companies). You can also run the free and open source Asterisk (http://www.asterisk.org) to have your own telephone exchange at home. (In addition, you can buy cards to plug into a computer running Asterisk into which you can plug regular phones or phone lines.)

MythPhone is in the standard set of MythTV plug-ins, so you probably have already installed it. Setting it up is relatively simple, but the error messages can be a bit lacking and leave you scratching your head.

MythPhone is not the be-all and end-all of Voice over IP (VoIP) applications. The current most-notable missing feature is echo cancellation. Many VoIP handsets (phones that talk VoIP protocols) or software have echo cancellation so that if there's enough latency on the connection and you're not using headphones, you can get annoying echo. This should not be the case just talking over a local area network (LAN), though. So, calling others in your house (for example telling the kids to come for dinner) should work brilliantly.

Getting a VoIP Account

There are many online resources for finding local VoIP providers. Also, if you often call a particular country, it's possible to get an account with a local provider there and just pay local call costs. There are two common types of accounts: a dial-out only account and a point-of-presence (PoP) account. Dial-out only accounts just let you make phone calls. PoP accounts also give you an incoming phone number. Many providers can give you a PoP phone number that's in one of many different areas that might or might not be anywhere near where you are. You can use this to your advantage—get a local phone number in another city so that people calling you from there pay only local rates.

You can also use Internet-only accounts that let you call other VoIP phones (and even toll-free numbers in many countries). One such service is Free World Dialup (FWD, http://www.freeworlddialup.com). It also has pay-for services, and as always, it pays to shop around.

The important information you'll need from your provider if you choose to have one is a user-name, password, and server name. You can happily use MythPhone without a VoIP provider if you just want to call other SIP-compliant devices on your local network or the Internet. A provider just gives you a user@host address, which means you don't have to somehow tell your IP address to the person who wants to call you.

Installing MythPhone

MythPhone is a MythTV plug-in that comes with the standard MythTV plug-ins package that you installed in Chapter 9. If for any reason you didn't install the plug-ins, you will need to install at least the MythPhone plug-in before continuing with this project.

Setting Up MythPhone

From the MythTV frontend, select Utilities / Setup ➤ Setup ➤ Phone Settings (as shown in Figure 14-1). You'll now see the first configuration screen for MythPhone (as shown in Figure 14-2). If you don't have this screen, it's because you haven't installed MythPhone yet—go to Chapter 9, and install the plug-ins before continuing with this chapter.

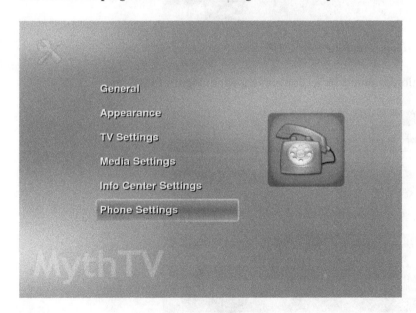

Figure 14-1. *Selecting Phone Settings in the frontend setup menu*

Next you need to configure the login information for your SIP provider if you have one. You can see an example of the login screen in Figure 14-2.

If you have an account with a SIP provider (such as Free World Dialup), make sure Login to a SIP Server is checked. If you don't have an account, make sure it's not checked (otherwise the MythTV frontend will take longer to start). The SIP Server DNS Name option is the machine to register with. The Sign-in Name option is often your phone number (or the user part of user@host, usually a number). The display name is just like how you enter your name in your email program—it exists only to look nice to the person you're calling.

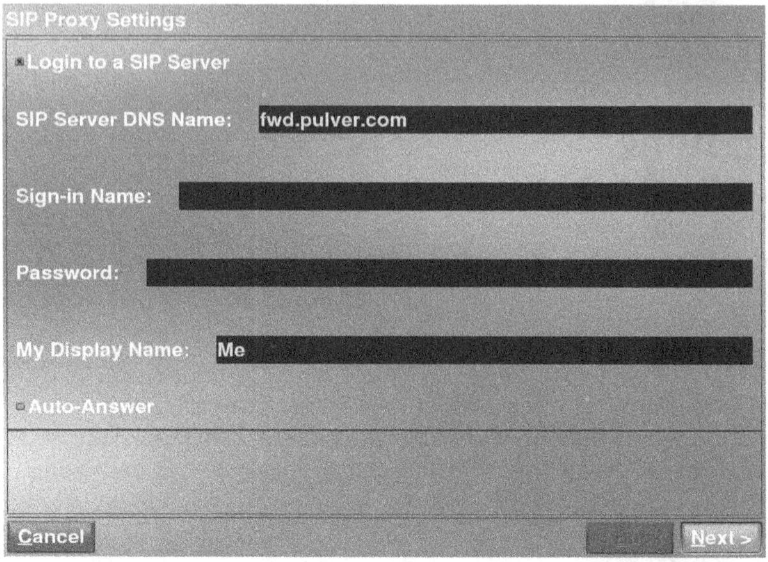

Figure 14-2. *SIP proxy and login settings*

The next configuration screen is IP Settings (as shown in Figure 14-3).

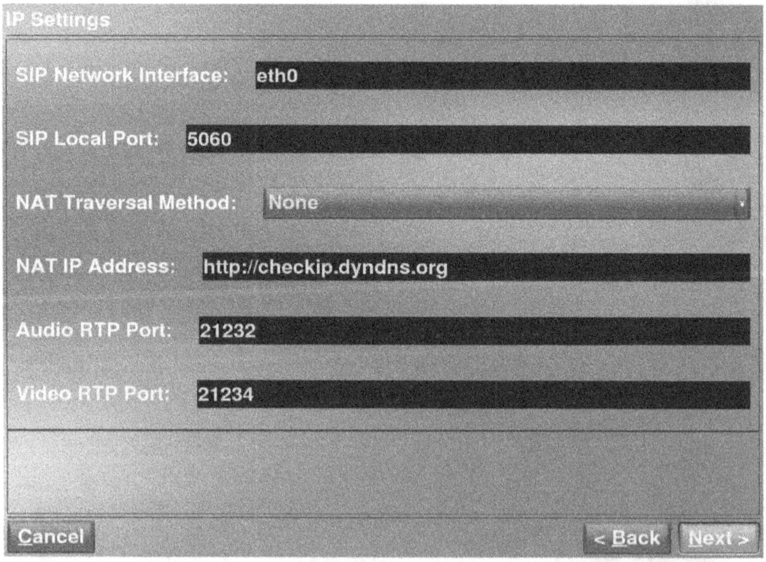

Figure 14-3. *IP settings for MythPhone*

These settings need to be correct to be able to receive any calls. The SIP Network Interface option is the network interface that is used to receive calls. The Network Administration tool (available through the System menu on the desktop) will tell you which network interface is which IP address (usually you will have only one, called eth0, but you might have more). You can also find out this information by running ifconfig from a terminal.

The SIP Local Port option is the port number that MythPhone will listen on for incoming calls. If during a call only one party can hear the other, you might have this port firewalled between the two parties. If you want to be able to receive calls from the Internet, make sure your firewall is set up for port forwarding of this port to your MythTV machine. With many popular gateways and routers (both wireless and wired) marketed for the home, you can easily do this through a web interface on the router; the specifics vary from model to model and from manufacturer to manufacturer.

The NAT Traversal Method option is the option to use to find out the real IP address of the MythTV machine if it's behind a network address translation (NAT) firewall. Most people now connect to the Internet through one of these types of networks. If you are just going to call local machines, None is fine. Select Manual if you always get the same IP address from your ISP (a static IP). There's also the Web Server method, which uses a web server out on the Internet to help find out your real IP address.

The Audio RTP and Video RTP ports are used for the audio and video streams. These UDP ports will need to be forwarded to your MythTV machine through your firewall. If you have multiple machines, they (naturally) cannot all use the same port through the firewall.

You will now need to set up some audio settings (as shown in Figure 14-4). The most important is the Microphone Device option. Note that if you use an analog TV card that has an audio cable that plugs into your sound card, you might have to get another microphone (for example, a USB one) to get sound that isn't from a TV station. The default values for the codecs and jitter buffers should be OK—modify them only if you know what you're doing or are having problems with sound or video "jittering."

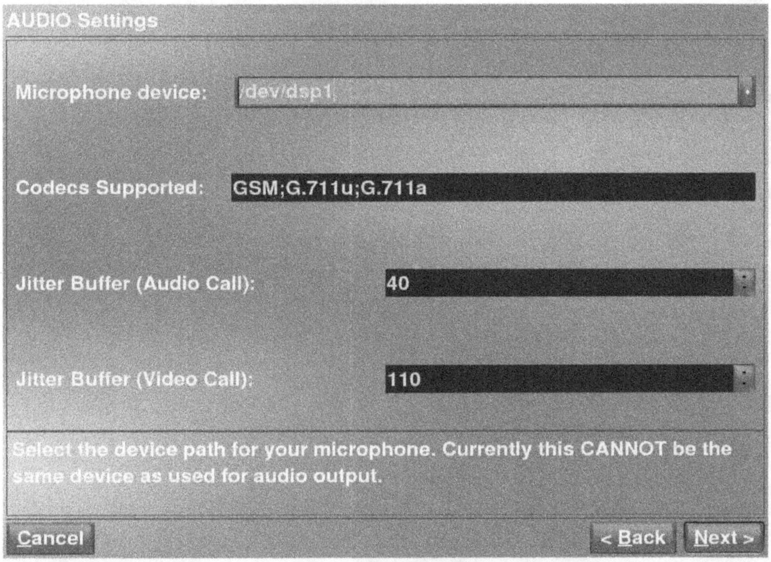

Figure 14-4. *Audio settings for MythPhone*

Next you can set up your webcam (as shown in Figure 14-5). Make sure you select your web-cam device. Most webcams present themselves as another Video4Linux device (just like your TV card), so make sure you select it and not your TV card. If you select the wrong one, you'll see the wrong image when you select Phone from the Information Center window from the frontend.

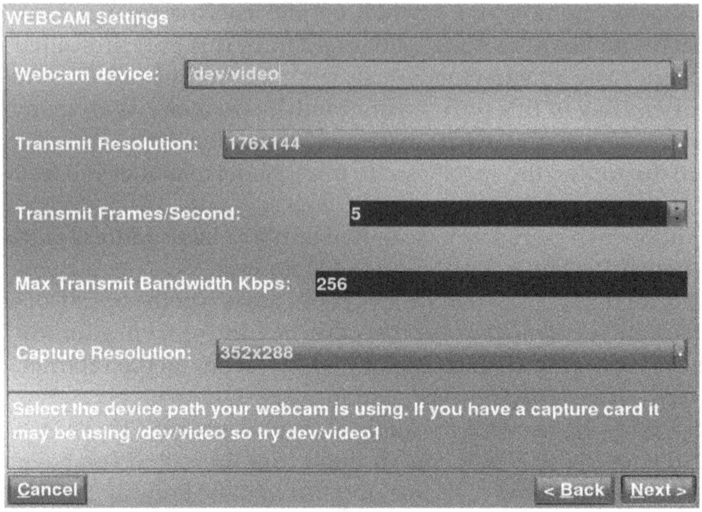

Figure 14-5. *Webcam setup for MythPhone*

If you're going to be doing video over the Internet, having a high Transmit Resolution setting probably isn't a good idea unless you have a very fast Internet connection. You can tweak the res-olution, bandwidth, frames/second, and capture resolution for your specific network setup.

VXML is a more advanced setup for some features that aren't often used (see Figure 14-6). Most users won't (and shouldn't) have to worry about it.

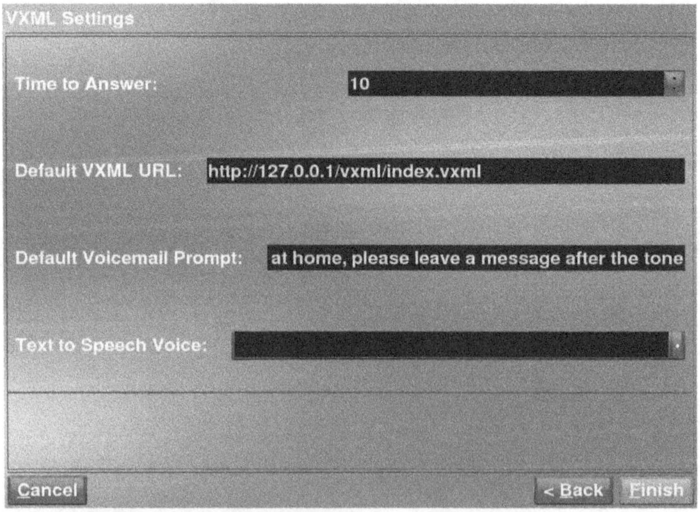

Figure 14-6. *VXML setup for MythPhone*

Now, after restarting the frontend, you should be able to select Phone from the Information Center window to start MythPhone.

In Figure 14-7, you can see me (Stewart) through my own webcam (hiding behind a camera) with the text in the blue status bar showing there are no active calls. The menu list shows two speed-dial entries: one for the current frontend and one for the other frontend on a machine called orpheus. MythPhone automatically creates speed-dial entries for each frontend with MythPhone set up on it. You can also press M or the menu key on the remote control to edit the directory (see Figure 14-8).

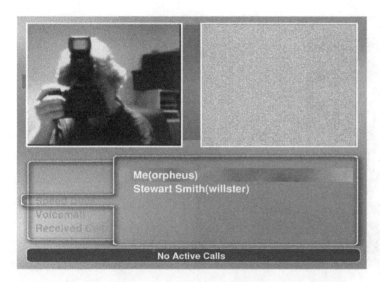

Figure 14-7. *Using MythPhone (not yet talking to anybody)*

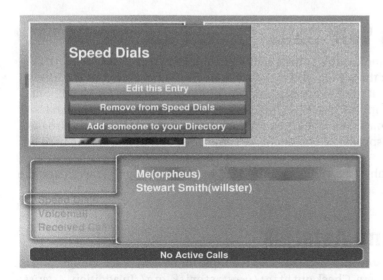

Figure 14-8. *The MythPhone Speed Dials menu*

In Figure 14-9 we've set up a directory entry for the talking clock number on Free World Dialup. The text entry for the URL can all be done with the number keys on your remote control, much like you enter numbers on your cellular phone. Since MythPhone automatically adds entries for each frontend with MythPhone set up, we can call our other frontend. Since we're also using a webcam, we can also have a video call to that frontend. By adding other SIP addresses (or using a SIP provider that allows dialing the telephone network), we can call other people comfortably from our living room. Dialing an entry is as easy as selecting it, pressing Enter, and selecting the type of call (video, voice, or instant message).

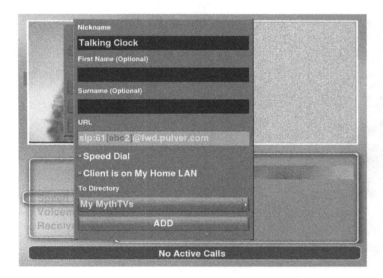

Figure 14-9. *Adding an entry to the MythPhone directory (the address book)*

Troubleshooting VoIP Calls

If you aren't getting very far, look at the output on the terminal when you run the MythTV frontend—specifically look for the SIP registration information. You should see your IP address in the output; if you don't, you'll need to tweak the network interface settings for MythPhone.

You can also perform a loopback test; to do this, press L, and both the local socket and NAT device should result in a call that echoes what you say back at you. Several SIP servers also have a number you can call for the speaking clock, a tone, or some other automated system used for testing purposes.

You can also tweak a number of settings (volume, FPS, resolution) by pressing F11 and the arrow keys.

Further Fun: Using Asterisk

You can set up your own PBX software: Asterisk. This is a topic for its own book, though, and is too vast a topic to be covered here. Check out http://www.asterisk.org/. In addition, *Asterisk: The Future Of Telephony* is a comprehensive book on the program, available in print and on the Web at http://www.asteriskdocs.org/modules/tinycontent/index.php?id=11.

Conclusion

In this chapter, we showed how to set up MythPhone to make calls to both SIP devices and (through a provider supporting it) calls to POTS (Plain Old Telephone Service). You learned that you can also use a webcam to get video calls to supported services. Although MythPhone has some limitations, you might find it a useful part of your MythTV setup—even if used only occasionally. It could be the start of much fun with other VoIP technologies.

CHAPTER 15

■■■

Joining the MythTV Community

This chapter focuses on something a little different from the others in this book. It's not specifically about getting a piece of functionality working in MythTV, although we will discuss how to get the latest version of the code from source control and run that. It's more about how to become an active member of the MythTV community, either as a developer or as a supporter of the community.

We will discuss how to install development releases of the MythTV code. We will also discuss the MythTV mailing lists that you should consider joining, the IRC channels devoted to MythTV discussions, and how to file bugs and keep track of the progress of MythTV development.

Joining the Users' Mailing List

The first step you should take when considering joining the MythTV community (or any other open source community) is to join the users' mailing list. You can find the MythTV users' mailing list subscription page at `http://www.mythtv.org/mailman/listinfo/mythtv-users/`. Be aware, though, that a fair bit of email flows through the list (it can be as much as 100 messages a day), so you should either set up mail filtering in your client or be prepared for all the email in some other way. And as with all mailing lists, it's a good idea to read before you write; join the list in advance and follow the traffic for a while so you don't make a fool of yourself in your first message—just when you need people to feel helpful toward you.

Chatting with Other MythTV Users

There are two MythTV-related IRC channels. (IRC stands for *Internet relay chat* and is an online form of "chat room" that predates both instant messaging and web forums. It is very popular in the open source community.) In IRC terminology, a chat room is called a *channel*.

To connect to an IRC channel, you will need to install an IRC client. We use Gaim, which you can install like this:

```
$ sudo apt-get install gaim
```

Once installed, you will need to configure an account. You do this by running Gaim from the Applications ➤ Internet ➤ Gaim Internet Messenger menu of the Gnome panel in Ubuntu. You will see the screen in Figure 15-1.

Figure 15-1. *Starting Gaim*

You need to create an account. Click the Accounts window in the background, and click Add, as shown in Figure 15-2.

Figure 15-2. *Creating an account*

You can see here that we want our username to be "newbie" (you can pick pretty much anything, and there isn't any need to register in advance), we're connecting to the Freenode

IRC network, and we want to log in when Gaim starts. Click Save, and you will see the screen shown in Figure 15-3.

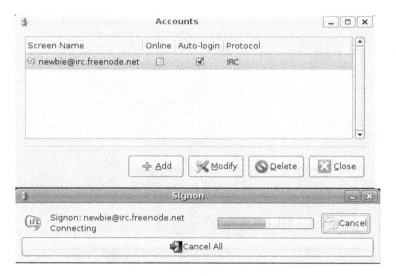

Figure 15-3. *Connecting to the IRC network*

You need to click the Online checkbox for the IRC account to connect to the IRC network. You need to do this only the first time, because next time the auto-start-up preference will be used. You can see the connection progress dialog box in Figure 15-3 as well. Once connected, you will be greeted with the screen shown in Figure 15-4.

Figure 15-4. *We're connected to the IRC network.*

Here we got an error message saying someone is using the "newbie" name that we wanted. That's OK; a different one will be issued to us if we don't authenticate as that user. You're now ready to connect to IRC channels. The relevant ones are #mythtv-users and #mythtv. Join a channel by using the join command like this:

```
/join #mythtv-users
```

A new tab will appear with the #mythtv-users channel (see Figure 15-5 for an example). This channel is used for the general user discussion of MythTV, such as helping others with their installations or asking installation questions. You should ask general questions here.

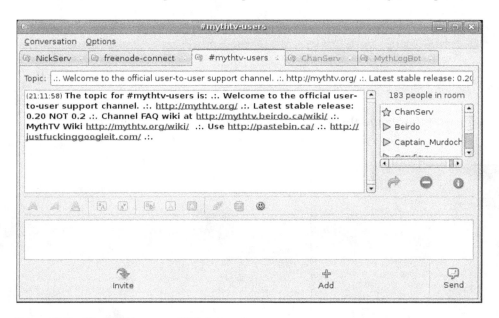

Figure 15-5. *The* #myth-users *IRC channel*

The second channel is #mythtv and is devoted to developer discussion. You can join in the same way as for #mythtv-users, but with the different channel name. Again, a new tab will appear. The #mythtv channel is for developer discussion only. Don't ask user questions here, because you will get a rude response! And note that, unlike in some projects, merely the fact that you're compiling the package from source code—even if on an unusual platform or operating system—doesn't qualify you to be asking questions on the developer channel; there are *lots* of MythTV users and not many developers, so preserving their peace and quiet is even more important than usual.

Finally, you can add these channels to your buddy list in Gaim for easy access later by clicking the Add button, which will present you with a dialog box like the one shown in Figure 15-6.

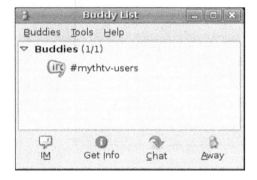

Figure 15-6. *The* #myth-users *IRC channel*

Now you have a shortcut to the #myth-users channel in your buddy list, as shown in Figure 15-7.

Figure 15-7. *The* #myth-users *IRC channel*

Helping Others

It is likely that other people in your area will also be trying MythTV. Often there will be several people at your local Linux user group (LUG) who will discuss MythTV-related topics every so often; there are Myth user groups in some cities as well. There are also commonly people new to MythTV. Sharing your experiences and learning from others is very rewarding, and we encourage you to do so.

Running the Latest Version of the Code

The MythTV developers change the MythTV code a lot. Most of those changes are incremental and aren't intended individually to create complete new versions of MythTV. Every now and then the developers release a version of these changes, and they try to make sure the release is as stable as possible. The developers then go back to breaking the code in order to make it even better for the next release.

This is a pretty standard pattern for the software development community, but in the open source community these incremental releases are public and reasonably frequent. The objective of making them available is so people will download them and give the changes a try and report any problems they experience. Additionally, it allows other people to develop an interest in the MythTV code and perhaps start contributing changes themselves.

You can find the Subversion (also called SVN, after the name of the command-line client) source code repository at http://svn.mythtv.org. You check out the latest code with the following command lines (first you'll want to create a directory specifically for building the software and cd to there first):

```
$ svn co http://svn.mythtv.org/svn/trunk/mythtv
A    mythtv/libs/libavcodec/qtrle.c
A    mythtv/libs/libavcodec/truemotion2.c
A    mythtv/libs/libavcodec/allcodecs.c
A    mythtv/libs/libavcodec/dsputil.h
A    mythtv/libs/libavcodec/error_resilience.c
...
A    mythtv/filters/denoise3d
A    mythtv/filters/denoise3d/filter_denoise3d.c
A    mythtv/filters/denoise3d/denoise3d.pro
A    mythtv/COPYING
 U   mythtv
Checked out revision 12195.
$ svn co http://svn.mythtv.org/svn/trunk/mythplugins
A    mythplugins/mythdvd
A    mythplugins/mythdvd/README-database
A    mythplugins/mythdvd/i18n
A    mythplugins/mythdvd/i18n/mythdvd_es.qm
A    mythplugins/mythdvd/i18n/mythdvd_nb.ts
A    mythplugins/mythdvd/i18n/mythdvd_si.qm
A    mythplugins/mythdvd/i18n/mythdvd_fr.ts
A    mythplugins/mythdvd/i18n/mythdvd_et.ts
...
A    mythplugins/mythnews/mythnews/mythnewsconfig.h
A    mythplugins/mythnews/mythnews/mythnews.pro
 U   mythplugins
Checked out revision 12195.
$ svn co http://svn.mythtv.org/svn/trunk/myththemes
A    myththemes/configure
A    myththemes/MythCenter
A    myththemes/MythCenter/preview.jpg
```

```
A    myththemes/MythCenter/theme.xml
A    myththemes/MythCenter/title
A    myththemes/MythCenter/title/title_info_center.png
A    myththemes/MythCenter/title/title_schedule.png
A    myththemes/MythCenter/title/title_games.png
...
A    myththemes/Retro-OSD/cut_end.png
A    myththemes/Retro-OSD/edit.png
A    myththemes/Retro-OSD/paused_frame.png
 U   myththemes
Checked out revision 12195.
```

Note those revision numbers: you'll need them if you have to discuss the code with other programmers. Once you have the code, the actual process of building and installing it is the same as in Chapter 3. We'll briefly repeat the relevant portions here, but refer to that chapter for a more complete discussion. First, run the configuration script:

```
$ cd mythtv
$ ./configure
# Basic Settings
Compile type     release
Compiler cache   yes, using ccache symlinked ccache distcc i486-linux-gnu-gcc-4.0
DistCC           yes, using distcc symlinked ccache distcc i486-linux-gnu-gcc-4.0
Install prefix   /usr/local
CPU              x86 (model name        : Intel(R) Pentium(R) M processor 2.13GHz)
Big Endian       no
MMX enabled      yes
Vector Builtins  yes
...
```

Next, compile:

```
$ qmake mythtv.pro
$ make
```

Lots of output will now be printed to the screen. This is building each part of MythTV. How long this takes will vary based on the speed of your machine, the number of processors in your machine, and the speed of your disks. It took around 15 minutes on one of our machines. After it has completed, you can install MythTV by executing make install with the appropriate permissions—this means running make install with sudo, like this:

```
$ sudo make install
```

You may be prompted for your password. All the files needed by MythTV will now be copied into directories under /usr/local/. We described how to install the plug-ins in Chapter 9. Here's a brief recap:

```
$ sudo apt-get builddep mythplugins
$ ./configure --enable-transcode --enable-vcd --enable-aac --enable-festival
```

Configuration settings:

```
        MythBrowser   plugin will be built
        MythControls  plugin will be built
        MythFlix      plugin will be built
        MythDVD       plugin will be built
        MythGallery   plugin will be built
        MythGame      plugin will be built
...
$ qmake mythplugins.pro
$ make
...
$ sudo make install
...
```

Next time you start `mythfrontend` on the machine that you ran `make install` on, the additional plug-in functionality will be available. Finally, you need to install the themes you just downloaded. Again, it's a case of following the `configure`, `make`, `make install` pattern:

```
$ ./configure
$ qmake myththemes.pro
$ sudo make install
```

Again, next time you start `mythfrontend`, the new themes will be available.

Updating the Source Code

Now that you have the source code downloaded, you can update your snapshot at any time by changing into the directory containing the code and running this command:

```
$ svn update
```

This will work for all of the three source directories you downloaded earlier. After you update, you can recompile using the steps listed previously.

Things You Should Know While Running the Latest Version

Running unreleased versions of any piece of software has risks. This is true of MythTV as well—you might miss out on recordings, you might corrupt your database, or you might lose data in some other way. The developers of course try hard to avoid these situations, but you should be aware that they might exist. If you need access to the cool, unreleased features available from the source repository, then remember to keep good backups and be careful about when you upgrade. You can find more information about backing up MythTV at `http://www.mythtv.org/docs/mythtv-HOWTO-23.html#ss23.5`.

Specifically, you should subscribe to the mythtv-commits mailing list, which is updated every time the code changes. You can find the web page for the list at `http://www.mythtv.org/mailman/listinfo/mythtv-commits/`. Here's an example of a commit message:

```
    Author: skamithi
      Date: 2006-12-06 07:26:35 +0000 (Wed, 06 Dec 2006)
New Revision: 12209
  Changeset: http://cvs.mythtv.org/trac/changeset/12209
```

Modified:

```
    trunk/mythtv/libs/libmythdvdnav/dvdnav.h
    trunk/mythtv/libs/libmythdvdnav/searching.c
    trunk/mythtv/libs/libmythdvdnav/vm.c
    trunk/mythtv/libs/libmythdvdnav/vm.h
    trunk/mythtv/libs/libmythtv/DVDRingBuffer.cpp
    trunk/mythtv/libs/libmythtv/DVDRingBuffer.h
```

Log:

```
Closes #2542. this should fix the madagascar dvd menu selection issue.
basically check if the title menu exists, if not, then don't use it.
i'm closing #2542 even though problem (1), the "double" button issue,
of the ticket is unresolved.
I don't know how to fix it, so if someone has a patch for it, please
reopen the ticket and submit the patch.
```

You can see in this message that the change was fairly safe and might be of interest if you've been trying to get that *Madagascar* DVD working. You should also join the developers' mailing list if you're interested in the general direction of the code and what people are thinking about at the moment. You can find the mailing list subscription page at http://www.mythtv.org/mailman/listinfo/mythtv-dev/.

Submitting Code

If you end up with changes to the code that you want to submit to the community, then you should follow the instructions on the MythTV website at http://svn.mythtv.org/trac/wiki/TicketHowTo. Basically, it's a case of filing a ticket instead of sending email to the developers' list. That way it's tracked and doesn't get forgotten.

Conclusion

This chapter was pretty short, but it should have set you up to become a more active member of the MythTV community. If you're looking for places to start helping the MythTV team out, we recommend working on some online documentation or the wiki (http://www.mythtv.org/wiki/index.php/Main_Page), hanging out on the mailing lists and asking questions, or playing with the code. What you do will of course depend on your free time and your interests.

Index

Symbols and Numbers

#mythtv channel (developers), 338
#mythtv-users channel, 338–339
>> (double greater-than signs), Xine and, 60
16:9 vs. 4:3 TVs for testing, 181–182

A

address book, MythPhone, 333
adjust audio sync option, 170
adjust time stretch option, 170
aggregators, defined, 209
alert template, 305
alert_text argument, 304
analog cable TV cards, 15
analog capture cards
    installing and testing, 43
    video capture cards, 17–18
angle brackets, XML, 185
appearance settings, 179
Apple remote control, 202
applications. *See also* mythfrontend
    dependencies, nuvexport command-line
        application, 142–146
    DVD player, 280–281
    Gaim, instant message application, 320
    transcoding, 282
    Yahoo! Go PVR application, 5
apt-get program
    as installation prerequisite, 48
    XMLTV installation and, 93
architecture
    of MythTV, 12–13
    xv acceleration architecture, 14
archiving. *See also* MythArchive
    DVD tracks, selecting for, 287–288
    DVDs, native archives export option, 300
    Gnome Archive Manager, configuring lirc
        daemon and, 57
aspect ratio option
    changing, 169
    custom themes and, 181
Asterisk PBX software, 327, 333
Asterisk: The Future of Telephony, 333
ATI
    aticonfig program, 45
    drivers, TV-out and, 45
    graphics cards, 13

audio
    Audio RTP, 330
    audio sync option, adjusting, 170
    settings for MythPhone, 330
Australian TV guide data, 92
authentication
    authenticate method, 313
    user, 316–317
autoexpire
    deleting recordings automatically and,
        127–128
    features of, 15
    list of recordings, 224–225
    turning off, 169
automatic login, 34–37
auto-skip commercials, 168–169

B

backends
    backend logs, checking (MythWeb), 274
    backend queue, 222–223
    backend status, checking (MythWeb),
        272–274
    configuring, 190–191
    defined, 12
    IP address settings and, 68
    starting, 80–82
background image, changing, 183–185
bandwidth, disks
    determining, 250–251
    multiple disks and, 252
batteries, remotes and, 56
bcastaddr command-line option, 304
begin transcoding options, 167–168
*Beginning Ubuntu Linux: From Novice to
    Professional* (Apress, 2006, 2007), 24,
    189
benchmarks, running on disk array, 250–251
Biggs, Billy, 181
binary-only ATI drivers, 14
bit rate, disk space and, 121
Blender program for 30 graphics, 187
blootube theme, example, 180
blue theme, 176
Bluetooth, wireless keyboards, 50–51
bookmarks, setting up in MythBrowser, 216
bootloader, reinstalling, 256

Find it faster at http://superindex.apress.com/

GPSR Compliance
The European Union's (EU) General Product Safety Regulation (GPSR) is a set
of rules that requires consumer products to be safe and our obligations to
ensure this.

If you have any concerns about our products, you can contact us on

ProductSafety@springernature.com

In case Publisher is established outside the EU, the EU authorized
representative is:

Springer Nature Customer Service Center GmbH
Europaplatz 3
69115 Heidelberg, Germany

www.ingramcontent.com/pod-product-compliance
Lightning Source LLC
Chambersburg PA
CBHW062050050326
40690CB00016B/3041

* 9 781590 597798 *